Hot Tech
Cold Steel

*How Computer-Aided Manufacturing Caught Fire
in Ann Arbor and Spread Around the World*

By

Charles S. Hutchins
Stephanie Kadel Taras

Produced by

TIME PIECES
PERSONAL BIOGRAPHIES, LLC

Ann Arbor, Michigan
Timepiecesbios.com

Cover design and layout by Ajuga, Inc.

One bee alone does not make honey.

—Ancient Greek proverb

CONTENTS

FOREWORD

It is an honor to write the foreword for this book that chronicles much of the professional life of my colleague and good friend Chuck Hutchins. I met Chuck early in my time serving as dean of engineering at the University of Michigan, and our association grew from there. Chuck was widely known throughout the College of Engineering and the broader university as a passionate supporter of Michigan, both financially and through his volunteer efforts where he so generously contributed his ideas and time. His enthusiasm and energy for all things Michigan were (and are) unmatched. In 2015, when a new football coach arrived and promised "enthusiasm unknown to mankind," he just did not know that Michigan already had Chuck. Chuck was not particularly interested in sports. Instead, he focused on the people of the university, especially its students, whether in the College of Engineering or the School of Music, Theatre & Dance. In the case of the former, Chuck became an early, ardent supporter of the University of Michigan Solar Car Team. He not only traveled regularly with the student team for multi-day competitions that traversed the United States and Australia, but he also offered insightful technical suggestions and hearty encouragement that contributed to their winning. It was in this connection that I came to recognize Chuck's engineering prowess, driven by his uncommon curiosity, vast experience in design and making, and familiarity with the latest advances in computing, materials, and other technical fields.

So, this is the man behind *Hot Tech Cold Steel*, which details the creation of the highly successful company Manufacturing Data Systems, Inc. (MDSI) that Chuck co-founded. MDSI pioneered the development of software for numerical control of machines used to create manufactured parts, but the story is much bigger. *Hot Tech Cold Steel* is about the manufacturing industry in the mid to late twentieth century, the early days of computing and networks, entrepreneurship and, most of all, about the people who created MDSI. Accordingly, this book is partly a history of the development of computing and its introduction into manufacturing, partly a guide on how to launch and scale up a successful start-up company, and also an informal treatise on the importance of relationships in identifying, hiring, and mentoring the right talent—everyone from precocious high school students to seasoned vice presidents. It is fair to say that the amazing team at MDSI, driven by their passion to integrate computing into manufacturing, anticipated cloud computing, software engineering, software as a service, personal computing, and even the global economy. Through shrewd business practices, MDSI was able to drive revenue by understanding the customer, delivering a revolutionary product that was sorely needed, and offering unparalleled service to its customers—a recipe for success in any business.

The reader will find that this book provides a clear exposition of numerical control of machine tools, with abundant technical detail, but with a narrative that is softened by story after story about the people of MDSI. This was a talented, motivated team that worked hard and played hard, under forward-looking company leadership. The MDSI culture was supportive and valued contributions from all. Of course, there were early hours, late hours, weekend hours, and enormous amounts of travel. There were times when money was running out. Life at MDSI was not always easy. But, it was hugely rewarding and fun, both personally and professionally.

One can only imagine that it was sometimes the families who were under the most stress, as their spouses, fathers, and mothers were totally consumed by the demands of the workplace at MDSI. In this regard, I would be remiss not to mention Ann, Chuck's partner and biggest supporter in life. In my mind, Ann represents the many unsung heroes who played a mostly silent role in enabling the creation of MDSI. It truly was a team effort. Cheers to the entire MDSI family, in recognition of your groundbreaking achievements!

David C. Munson Jr.
President
Rochester Institute of Technology

PREFACE

Was it really faster, easier, better, and cheaper? To get an answer to that question, Chuck Hutchins and Ken Stephanz were standing inside a manufacturing facility in Cleveland, Ohio, watching a young man punch codes into a teletype machine. It was 1968. There was no internet. So the young man used a dial-up phone line to connect to General Electric's (GE) timeshare computer service. That procedure resulted in a punched paper tape that ran the factory's machine tools.

Chuck Hutchins had developed a competitive service, a version of a programming language for numerically controlled (NC) machine tools (mills, lathes, drills) that he believed was better than what was offered by the behemoth General Electric. After months of tinkering and testing, with the help of two other young computer programmers, Chuck was ready to take this service to the marketplace. He had an arrangement with one of the country's few timesharing computer companies at that time—Comshare of Ann Arbor, Michigan—and he felt confident that, with the right investment and leadership, his new approach could grow into a successful business.

Chuck had only recently met Ken Stephanz in Ann Arbor, where they both lived. Chuck, a wiry guy in his mid-thirties who talked fast and laughed heartily, impressed Ken, an experienced corporate executive in his early forties whose low voice matched his calm, controlled demeanor. Ken was taken by Chuck's enthusiasm and obvious intelligence, but after their first meeting, he thought, *Who is this wild, messianic guy who believes he can change the numerical control machine tool manufacturing world single-handedly with his technology?* Ken was intrigued by the enterprise, but he needed to know if Chuck's idea was financially viable.

It was Ken who asked Chuck to demonstrate his program in a head-to-head comparison with the competition. It was Chuck who set up this test in Cleveland, with the understanding that the manufacturer would tell them how much GE was charging for its service.

The GE timesharing service and Chuck's new technology used the same basic process: type a series of coded instructions into the bulky teletype keyboard; connect via telephone line to an off-site computer that interpreted the code and created a 1"-wide, punched paper tape that was punched at the factory; feed the tape into a reader connected to the NC machine tool; and watch the tool perform the job with automated precision—cutting a part from metal, drilling a series of holes into the part, and so on.

After the process was completed using GE's timesharing computer, Chuck stepped up to the teletype machine to enter his code that would tell the machine tool to do the same function. Tracking the time needed to enter the code and the time required by the off-site, timesharing computer to process the instructions, Ken could already sense that Chuck's process was indeed faster and easier. And when the punched tape worked exactly as expected, Ken could see it was just as effective.

As they left the Cleveland plant to drive back to Ann Arbor, Ken began running the numbers in his head. He already knew that they would have to pay Comshare $.03 per CPU second for use of their timesharing computers. And now he knew how much GE was charging for their slower approach.

A smile spread across Ken's face. Turning to Chuck, he said, "Do you realize we could charge $.30 a CPU second and still be cheaper than GE's timesharing? We could charge *ten times* our cost and still be competitive!"

Chuck slapped his hands together, his eyes wide. "I knew it!"

Chuck had, in fact, created a service that was faster, easier, better, and cheaper. And Ken felt sure he could find investors for the profits he thought were possible.

■　■　■

Seven years and a few thousand customers later, the company co-founded by Chuck and Ken, Manufacturing Data Systems, Inc. (MDSI), would buy seventy-seven acres of land on Plymouth Road in Ann Arbor and would go on to build a 200,000-square-foot complex paid for with cash from revenue. With some 850 employees worldwide at its peak, MDSI was one of the largest private employers in town.

By the end of the 1970s, MDSI software would program almost twenty thousand numerically controlled (NC) machine tools, not only in the United States but in much of the world. In recognition of its widespread use, the American National Standards Institute designated MDSI's proprietary software as a National Standard. In addition, with wholly owned subsidiaries in Canada, England, France, Germany, Switzerland, Sweden, and Japan, "MDSI literally taught the world to program NC machine tools," says Chuck.

MDSI went public in 1976, and in 1985, *Forbes Magazine* designated it the eighth most successful Initial Public Offering of the top one hundred IPOs in the 1975–1985 decade. MDSI started with $1.5 million of venture capital invested in 1969 and 1970, and was sold in 1981 to Schlumberger for $210 million.

By every measure, Chuck Hutchins, Ken Stephanz, and the MDSI team hit a home run for industry, for their community, for their employees, and for their investors. And they did it with computers several years before Microsoft and Apple had even been conceived.

■ ■ ■

MDSI began at the key moment when computers leaped from the research lab to useful applications, when very few people understood what computers could do, let alone how to program them. The story of MDSI is a foundational story in the history of computing, in the evolution of automation, and in the annals of tech company fairy tales that would later be told about Apple, Google, and others.

In the years since MDSI's dominance, Ann Arbor has positioned itself as a hub for tech innovation in the Midwest. With the University of Michigan (U-M, or simply "Michigan") at the leading edge of related research from robotics to biosciences, with tech entrepreneurs driving advances in big data analysis and device security, and with a commitment to bringing the region's automotive past firmly into the self-driving future, Ann Arbor's business leaders are excited about the future. But they may not be aware that they are building on layers of tech leadership going back to the dawn of the computer age.

Perhaps the story of MDSI is not better known because it served the manufacturing sector rather than the wider public, or because it predated personal computing, or because it preceded the internet and the tech bubble by more than two decades. But any tech entrepreneur today will see that the MDSI story is rife with Silicon Valley tropes: the visionary engineer with a limited head for business, the venture capitalist looking for his first big score, the experienced CEO with his corner office and complex spreadsheets, the monomaniacal programmers, the high-tech corporate campus, the IPO, and the mad rush to hire followed later by massive layoffs. MDSI did it all in Ann Arbor, more than fifty years ago.

In the process, it drove one of the most significant changes in the world economy—factory automation.

In the years before founding MDSI, Chuck Hutchins personally brought NC machines from the rarified world of military-funded aerospace companies to Michigan's automotive companies and suppliers. He didn't invent NC machines, and MDSI wasn't the only company in the 1970s that taught the world how to run them. But Chuck's user-friendly software, combined with expert customer support, ambitious international expansion, and a synergistic team, captured the lion's share of the NC machine tool market at that time. For more than ten years, MDSI helped convert machine shops all over the world to numerical control.

Next, Chuck and his ever-curious software engineers used the resources of MDSI to push ever-advancing computer technology into these same shops. Their innovations helped manufacturers go from timesharing off-site computers to stand-alone on-site computers, from computer-aided manufacturing to computer-aided design, from two-dimensional digital drawings to three-dimensional digital drawings, from automated 3-axis machines to automated 5-axis machines. And once a factory was using computers to run its machines, there was no going back. The fully automated, robotic factory can be viewed as the ultimate outgrowth of MDSI's work.

At the same time, the engineers at MDSI advanced computer technology itself. They developed internal capabilities—including email and accounting software—that anticipated the corporate information age. They built early desktop computers that rivaled Apple's earliest inventions. They experimented with memory-saving coding solutions that only the nation's most elite computer programmers understood. Some of MDSI's software engineers went on to become leading experts in higher-level languages, microcoding, and supercomputers.

Taking MDSI's influence even further, it can be seen today in the technology of 3-D printers, which might be described as miniature, numerically controlled milling machines that squirt melted plastic rather than cut metal. And of course, the results of computer-aided design are ubiquitous in our lives today, from architecture to household products.

Any appreciation of the automation age and Ann Arbor's position in the digital revolution requires an understanding of MDSI's role in getting us here. And a full appreciation of that story starts in the machine shop long before computers arrived . . .

Part I
The Lead Up

1

The Engineer

Chuck Hutchins was fifteen years old in 1949 when he learned firsthand the value of automated manufacturing. He was working a summer job in Sam Gray's sheet metal shop in Pontiac, Michigan, when a job came in from the Pontiac High School Band for 125 sheet metal rings. The large marching band would be performing that fall at nighttime football games at Wisner Stadium, and the band director had the idea to attach a metal ring with nine battery-powered light bulbs onto the top of each band member's hat.

Sam handed Chuck 125 10" squares of sheet metal and told him to locate the nine holes for the light bulb sockets, along with another nine holes for the rubber grommet that would insulate the metal strip where it made contact with the bottom of the bulb.

Chuck remembers, "I located the center of the sheet and hit that spot with a center punch. Then I drew a 4.5"-radius circle around the center and punched the first socket hole. Then I carefully measured 40 degrees around the circle and punched a second spot. I quickly realized this was *not* the way to lay out 125 sheets, much less 1,125 holes. Even worse, twice that many!"

After considering the problem for a few minutes, Chuck came up with a simpler strategy: He put a .25" hole in the center of each 10"-square sheet of metal. Then he put a .25" pin in the punch press table 4.5" from the punch. He put the first piece of sheet metal on the pin and punched the first hole. He then took the slug from that punched hole and filed the edges so it would drop easily back into the hole. After spot-welding that slug to a long narrow finger, he rotated the 10"-square plate to the location for the second hole, positioned the finger with the slug into the first hole, and clamped it to the punch press table.

"Now, as fast as I could lift the finger, rotate the sheet, replace the finger, and hit the lever that actuated the punch, I went bang, bang, bang, for nine equidistant holes. Those 125 sheets went pretty fast!"

After rotating the sheet manually to the location of the first grommet hole, he clamped the finger to a new location on the punch press table and repeated the process for the second set of nine holes. He had punched all 2,250 holes in their precise locations before the end of the afternoon. Says Chuck, "I'm sure that early engineering experience played a significant part in leading me to engineering."

Even more significant was the next step in the process. As the summer went on, the company that was going to make the light bulb sockets reneged on their commitment. When Chuck heard this news, he saw an opportunity to make some more money. He said, "Sam, I've got a lathe at home. I could make those light bulb sockets." Sam was going to pay $.29 for each socket—$326.25 for 1,125 sockets. Says Chuck, "That was almost half of what I'd paid for the lathe! It made a fifteen-year-old kid's eyes light up."

Chuck had purchased the lathe two years earlier, at the urging of his great-uncle, Will Foltz—"Uncle Will." (Also called "Bill.") When Chuck was just ten years old, he started visiting Foltz's Royal Oak machine shop, on Ten Mile Road east of Woodward Avenue. Foltz, a former engineer for Cadillac, had started his own company making lead gaskets to seal up 55-gallon drums for chemical solvents. The machine shop captured Chuck's curiosity, and Foltz had no trouble teaching the boy how to run the World War I surplus 16" South Bend engine lathe to bore out these gaskets. In 1946, Uncle Will had suggested that if Chuck had a 9" South Bend lathe at home, he could pay Chuck to make special-sized nuts for his business.

Chuck remembers, "My dad and I sat down with the South Bend lathe catalog to figure out what I would need. The lathe, plus three chucks, the tool holders, and miscellaneous other parts added up to $750, a lot of money in 1946. Dad made me a deal: if I could earn half, he'd pay the other half." The snow was so bad that winter that schools in Pontiac were closed for more than a week. Chuck shoveled every sidewalk in the neighborhood, and by spring of 1947, he'd earned half the cost. "Dad and I went to the Lee Machinery Company on Jefferson Avenue in Detroit and placed the order. Thus, when I was thirteen years old, I owned my own 9" South Bend lathe."

For the next two years, Chuck used the lathe occasionally, but when he offered to make the light bulb sockets for Sam Gray, he wasn't fully aware of what he was getting himself into. "The first problem was making a tap for the rolled thread of the light bulb socket. I chased the thread on a piece of 3/8" drill rod, filed the roundness on the threads, then heated the tap with a torch and quenched it to make it hard, then ground the flutes. It was crude, but when I used the tap to put the threads in the hole, the light bulb screwed into the hole successfully. One thousand, one hundred, and twenty-five sockets later, I knew one thing: I did *not* want to be a lathe operator!"

If that realization made a college education more attractive to Chuck, these early experiences with manufacturing also highlighted the machine tool operator's endless quest for greater speed and efficiency. Could there have been a better background for a man who would one day apply computing to that very goal?

■ ■ ■

Growing up in Pontiac in the 1940s gave young Chuck many opportunities to see the thriving automotive industry in action. Both sides of Chuck's extended family were involved in automobile manufacture in different ways.

His mother's father, Ernest Edwin Sweet, came to Detroit about 1890 and got a job with Leland, Faulconer & Norton Co., which made machine tools, gears, and an internal combustion engine. Sweet was essentially a self-taught mechanical engineer, having learned engineering by the study of four textbooks purchased from the International Correspondence Schools of Scranton, Pennsylvania.[i]

Working closely with Henry Leland, Sweet was largely responsible for the design and construction of a ten-horsepower, single-cylinder internal combustion engine. His employer, which became Leland & Faulconer when Norton left in 1894, supplied the transmission for the "Curved Dash Oldsmobile"—the first automobile to be manufactured

---

i   Chuck still has these four textbooks.

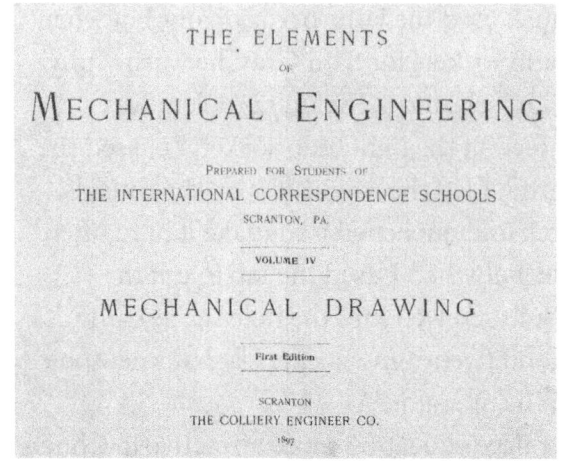

THE ELEMENTS
OF
MECHANICAL ENGINEERING

PREPARED FOR STUDENTS OF
THE INTERNATIONAL CORRESPONDENCE SCHOOLS
SCRANTON, PA.

VOLUME IV

MECHANICAL DRAWING

First Edition

SCRANTON
THE COLLIERY ENGINEER CO.
1897

Title page from one of Chuck's grandfather's textbooks that he used to teach himself mechanical engineering.

on an assembly line.[ii]

When Henry Ford Company (the precursor of Ford Motor Company) was disbanding in a dispute between Henry Ford and his investors, the financial backers hired Henry Leland to appraise some of their assets. The money men were impressed with Leland's advanced engine design, and Leland suggested, "Let's put that engine in the car and continue building automobiles." The result was Cadillac Motor Car Company, established in 1902. Ernest Edwin Sweet—Chuck's grandfather—became Cadillac's founding chief engineer.

Leland and Sweet were committed to quality, and Cadillac quickly became known for its precision machining, with parts that were interchangeable from one vehicle to another without any filing or sanding. In his 1926 autobiography, Leland wrote, "There always was and there always will be conflict between Good and Good Enough, and in opening up a new business or a new department one can count upon meeting this resistance to a high standard of workmanship. It is easy to get cooperation for mediocre work, but one must sweat blood for a chance to produce a superior product."[1]

When General Motors (GM) purchased Cadillac in 1909, Sweet continued his work there.

Decades later, Chuck had the opportunity to speak with a man who had worked with Sweet. Chuck remembers, "In 1983, I saw an announcement in the University of Michigan Engineering alumni news about Fenn Holden, a 1908 mechanical engineering graduate who had

---

ii Chuck is fully aware of the many ways history can be told. He points out that, in the early 1900s, his grandfather Ernest Sweet was a hero, because his development of that first Cadillac single cylinder engine (and its continuing refinement) eliminated the daily removal of tons of horse manure from the streets of every major city and town. Now, one hundred and ten years later, Sweet can be seen as a villain, as car engines have helped to pollute the environment to the point that humanity may not be able to survive in the atmosphere that will exist on the Earth one hundred years from now.

gone to work for Cadillac. I thought, I wonder if he knew my grandfather?" A few phone calls later, Chuck located Holden in Jacksonville, Florida. When Chuck went to visit him, Holden had just celebrated his one hundredth birthday, seventy-five years after graduating from Michigan.

The 1902 Cadillac. Picture taken at the Gilmore Car Museum in Hickory Corners, Michigan.

Chuck asked him, "Did you know a man named Sweet?"

"Ernest Sweet?" said Holden. "He was the *big cheese* when I started at Cadillac."

Chuck then asked, "Did you know a man by the name of Foltz?"—asking about his Uncle Will. Holden answered, "Bill Foltz? He was the foreman of four-cylinder engine assembly, and he was my first boss when I joined Cadillac."[iii]

Chuck then asked for stories about those early Cadillac days. Holden said that, in 1912, Henry Leland's son, Wilfred, came to talk with Ernest Sweet about an idea for a more powerful motor:

Wilfred said, "We build this four-cylinder engine," holding up four fingers of his left hand. "What if we put two together?" Wilfred then held up four fingers on his right hand, meshing his fingers into a V-shape. "Couldn't we put two together and make a V-8?"

The gasoline-powered V-8 engine had been invented in 1902, but it was Cadillac, under Sweet's leadership, that introduced the V-8 engine into mass-produced automobiles in 1915.

After the United States entered World War I, Leland, who was very patriotic, wanted Cadillac to build the Liberty aircraft engine for use

---

iii Fenn Holden was involved in the design and construction of the GM Proving Ground in Milford, Michigan. In particular, he designed two one-mile-long straightaways (that are named after him) at right angles to each other so that performance data could be gathered while compensating for the effects of blowing wind.

A 1907 Cadillac on the east shore of Cass Lake, three miles west of Pontiac, Michigan, in the summer of 1909. The little girl in the back is Chuck's mother, Virginia, seated with her four older sisters. Chuck's grandmother, Ida May Sweet, is driving.

in the war effort. But General Motors President Billy Durant refused, believing the U.S. had no business getting into the war and that the war would be short-lived. So, in July of 1917, Henry Leland; his son, Wilfred; and Ernest Sweet left Cadillac to start a new venture—Lincoln Motor Company. The first thing they made were 18,000 Liberty aircraft engines. And then they built the Lincoln automobile, which was bought in 1922 by Henry Ford.

Ernest Sweet had long been friends with Bill Foltz, going back to their earliest days as engineers at Leland & Faulconer. As young bachelors, they lived in a rooming house in Detroit owned by the Dewey family. The Deweys had two daughters, Ida May and Blanche, and before long, Ernest Sweet and Bill Foltz married the sisters. Ernest and Ida May Sweet had four daughters, including Virginia Belle Sweet (born in 1906), who would become Chuck's mother.

Virginia married Chauncey Sanborn "Hutch" Hutchins on May 2, 1928. Their first child was Charles "Chuck" Hutchins, born February 3, 1934. Chuck's sister, Julie, was born two years later. About that same

HOT TECH COLD STEEL

time, Ernest Sweet died. Says Chuck, "I don't remember my grandfather. But his career does make me wonder if engineering could possibly be an inherited genetic trait."

Since Bill Foltz was married to Ida May's sister, Bill was Chuck's great-uncle, and that's how Chuck came to be in Uncle Will Foltz's machine shop in Royal Oak in the 1940s to learn how to operate a lathe.

One other machine from Chuck's childhood may have influenced his eventual career. His mother's sister Ethel had a player piano—a 1926 Steinway Grand with a Duo-Art Player attachment. It operated with hole-punched scrolls of paper, much like the early computers to come, and Chuck loved to watch that scroll tell the keyboard how to play a song—without any assistance from a person.

■　■　■

Chuck's father, Chauncey Sanborn Hutchins, was born in Pontiac in 1904, the son of Chauncey Hubbard "C.H." Hutchins and Lottie Sanborn Hutchins. The elder Hutchins founded the Pontiac Varnish Company in 1902 in a garage in downtown Pontiac. About 1907 he bought property for his company at 30 Brush Street on the south side of Pontiac. It was just one mile from the General Motors truck plant, so it was well positioned to serve the Varnish Company's customers.

C.H. Hutchins—Chuck's paternal grandfather—was making varnish in big kettles. He melted pine pitch over huge oil burners, added "magic ingredients," and called it varnish. Says Chuck, "I have no idea where he got the pine pitch or what the magic ingredients were at that time."

Chuck's father grew up around the varnish factory. He graduated from the University of Michigan in 1927 and returned to Pontiac to join the family business, eventually becoming executive vice president.

After he married Virginia, they moved into an apartment house in Pontiac where their neighbors, a young couple named Harold and Frances Furlong, soon became their best friends. Harold, a Pontiac native who had won the Congressional Medal of Honor for his bravery in World War I, went to medical school at the University of Michigan, became an obstetrician, and after five years working at the university, felt obligated to return to Pontiac to care for the women in his home community.[iv]

The two couples continued to live near each other, even after moving to other homes in Pontiac. When Chuck was born, he learned to call Harold and Frances "Uncle Fuzz" and "Auntie Fran." One of his earliest memories is going into the backyard with his mother to meet up with Auntie Fran on the other side of the hedge that separated their yards. In 1939, when Chuck was five, his parents moved to a house on Ottawa Drive, where they lived through the war years.

During World War II, the Pontiac Varnish Company had contracts with the military to make, among other products, a wax coating for canvas army tents and the olive drab paint for GM 6 x 6 army trucks. Chuck remembers seeing the paint being shipped to the GM factory in tank wagons that drove a mile down Franklin Road to GM Truck. "But the thing I remember most about that time was every Sunday night we would take Dad to the train station in Detroit so he could go to Washington, D.C. And then we'd pick him up on Friday night. It wasn't every week, but in my memory, it was a lot." Chuck presumes his dad was meeting with government officials to keep the contracts coming.

Chuck's grandfather, C.H., owned a 175-acre farm near Pine Lake and Orchard Lake and saw himself as a "gentleman farmer" (who hired others to do the actual farming). During the war, in the field behind the farmhouse, C.H. created a family Victory Garden with 100-yard-long rows. "I can remember planting potatoes," says Chuck, "duck walking with a five-gallon pail down this long row. By the time I got to the other end, I could hardly stand up. But our whole family got all our vegetables during the war from that garden, and my mother and my two aunts canned the produce for the winter."

When the war ended, Chuck's father had more time to spend with his son, but their relationship was cut short when Hutch died on August 18, 1947, before he reached his forty-third birthday. Bronchial pneumonia was the official cause of death, but smoking was the culprit.

---

iv Harold Furlong eventually delivered more than 4,000 babies in the Pontiac area, and even delivered Ann and Chuck Hutchins's first daughter, Linda. Also, curiously, while he was a resident OB-GYN at Michigan, Dr. Furlong delivered the doctor who would one day deliver Ann and Chuck's second daughter, Beth.

HOT TECH COLD STEEL

"My father smoked three packs of Philip Morris unfiltered cigarettes every day for twenty years," says Chuck. "That's sixty cigarettes a day. I figured out that, if he was awake from 7:00 a.m. to 11:00 p.m., he had to light a cigarette every fifteen minutes of his adult life."

The funeral was one of the biggest Pontiac had ever seen, with a procession of cars that stretched more than a mile down Woodward Avenue. As Chuck remembers it, many attendees were the parents of his friends, who lived in a tight-knit neighborhood.

Chuck still gets choked up remembering that terrible loss when he was just thirteen years old. It was only a few months after his dad had helped him buy the lathe. "He never knew the impact of that investment." Chuck can't imagine how distraught his mother must have been. His parents had not yet been married twenty years—an unfathomably short time to Chuck, who has now been married to Ann more than three times as long.

With C.H. retired and Hutch deceased, Chuck's uncle Paul Ziegelbaur took over as president of the Pontiac Varnish Company.

■ ■ ■

In an effort to bring male influence into Chuck's life following his father's death, Chuck's mother immediately enrolled him at Cranbrook School for Boys. The private, college prep school in Bloomfield Hills was twenty years old at that point. The faculty were all men (except the typing teacher). The stunning campus (designed by the distinguished architect Eliel Saarinen) had a dormitory for boarding students, but Chuck lived at home, taking the school bus about thirteen miles via a circuitous route to Cranbrook. In his senior year, his paternal grandfather, C.H., gave Virginia the house on the farm on West Long Lake Road; after she moved there with her two children, Chuck lived closer to Cranbrook.

"My years at Cranbrook made a real difference for me," says Chuck. He had two especially good teachers who became lifelong friends. Dick Hintermeister, the shop and drafting teacher, taught Chuck a range of woodworking skills, while Floyd Bunt taught him chemistry and

Chuck Hutchins as a student at Cranbrook School for Boys, 1952.

supervised his work in the "auto lab." In the years following World War II, as car companies were ramping up production, they built auto labs in area high schools to teach students how to work with automobile engines. "For five years, I spent the majority of my non-class time in the shop or the auto lab." In addition, Chuck helped with stage lighting for the school's Gilbert and Sullivan productions.

In 1949, one of Chuck's older cousins gave him a beat-up 1931 Chrysler eight-cylinder Roadster with a rumble seat. Chuck remembers, "It was parked in a barn, and he said it was mine if I could make it run. I believe he thought I needed something to keep me out of trouble." After many trips to a gas station to recharge the two batteries he had brought to the barn, Chuck got the car started and drove it home. "But I found it didn't have enough power to shift into high gear. Floyd Bunt coached me through rebuilding that engine from a bare cylinder block." Chuck drove that car for the next six years, including two trips back and forth to Troy, New York, while he attended Rensselaer Polytechnic Institute (RPI). "I had that car until I bought a new 1955 Chevy for $1,829—taxes, license, and title included."

Chuck graduated from Cranbrook in 1952 in a class of just fifty-two young men. "I had been in the Class of '51, but at the end of that school year, my mother and the headmaster, Harry Hoey, thought I was too young to go away to college, so I stayed at Cranbrook for an extra year. I don't remember if I was even consulted."

By then, he knew he wanted to be a mechanical engineer. "I'd probably known that since I was ten years old." Next to his photo in the 1952 Cranbrook School yearbook is this description: "Charlie just could not be happy if there were no motors in the world. A natural born tinkerer, he finds a way to spend as much time as possible either working on automobiles, fixing stage lights, or doing just about anything that requires mechanical ability. As one might suspect, he is planning on an engineering career."

The University of Michigan was the obvious choice for a budding engineer from Pontiac, but Chuck wanted to study at a small school. He spent the next two years at RPI. But it was a long way from home, and the tuition was significantly higher than at Michigan. "I finally realized that, even though Michigan was a big school, the College of Engineering was similar in size to RPI. And tuition was only ninety dollars a semester."

In the fall of 1954, he transferred to Michigan, thus changing the entire trajectory of his life.

■   ■   ■

The University of Michigan's North Campus was just being built when Chuck arrived, and the College of Engineering was still located in the Engineering Buildings on Central Campus, on the corner of East University and South University. Chuck still had three years to go to meet all the course requirements for an engineering degree at Michigan, and while he just wanted to do hands-on lab work, he still had several basic courses to complete.

One challenge for Chuck was the required course in calculus. "I wasn't a particularly good student," he says, "and calculus just went right over my head." He failed the class. "Before I knew it, I'd received a letter from the dean suggesting I should continue my education somewhere else."

When Chuck's mother saw the letter, she phoned Harold Furlong for help. In the years after Chuck's dad died, "Uncle Fuzz"—the obstetrician with a medical degree from Michigan—served as a kind of "pseudo-dad" for young Chuck. One Saturday afternoon, Dr. Furlong called Chuck over to his house for what turned out to be a seminal conversation in Chuck's life. "He chewed me out. Today, I call it my 'Dutch uncle' lecture."

After giving Chuck a good talking to about studying more and taking things more seriously, Furlong asked, "Now, what are you going to do on Monday?"

Chuck said, "I'm going to Ann Arbor."

"What are you going to do in Ann Arbor?"

"I'm going to see the dean."

"What are you going to say to the dean?"

"I'm going to ask him, 'What do I have to do to get back into Michigan?'"

"Then what are you going to do?"

With tears running down his cheeks, Chuck said, "Uncle Fuzz, I'll do anything he tells me to do."

That's exactly what happened. With the dean's encouragement, Chuck enrolled in summer school at Wayne University and got a B average in two classes—Calculus and Strength of Materials. That got him back into the Engineering program for the fall of 1955, which was his junior year. "I wouldn't have a University of Michigan degree or a whole lot of other things," says Chuck, "if it hadn't been for Harold Furlong."

Among the courses Chuck took in his junior year was an elective in sociology called "Marriage Relations." It was the spring semester in 1956. "The professor assigned us to write a paper," Chuck remembers, "and then he dropped a bomb on us: 'I want the paper written in groups of two.'" Chuck knew no one in the class. He looked around, and his eyes landed on a tall, young woman with a gorgeous head of dark, naturally curly hair. Her name was Ann Strong. Chuck soon learned Ann was from Oxford, Michigan, and a student in the School of Art & Design. The two classmates wrote the paper together. Later, Chuck went to the School of Art & Design to see what Ann was doing. "Wow, was she talented!" he says. Soon a romance blossomed. Chuck and Ann were married on September 1, 1956, Labor Day weekend, before they started their last year at U-M.

In his senior year, Chuck could finally focus on the classes he really wanted to take. He enrolled in the first ever class taught in U-M's new Automotive Engineering Laboratory on North Campus. "I was one of a very few students, perhaps the only one, who had rebuilt an internal combustion engine before taking that course."

Chuck also signed up for Elements of Machine Design with Professor Herbert Alvord, and then Advanced Machine Design with Professor Keith Hall. "These courses were at the top of my list," says Chuck, "the things I'd wanted to study from day one."

Alvord had come to Michigan after several years as a practicing machine designer, first with Fairbanks-Morse in Beloit, Wisconsin, and then with Cummins Engine Company in Columbus, Indiana. That experience impressed Chuck. "I'll never forget my first day in his class. He said something like 'I'll expect you to read the textbook. In class, I'll answer any questions you may have. But most of class time will be spent describing and discussing real machine design problems and solutions that I have experienced at Fairbanks-Morse and Cummins Engine.' " [v]

The next semester, in Advanced Machine Design, students were assigned a real-life task: design a three-station press to make a knob for a steam radiator valve. Most of the students chose to make their presses with cast iron frames, but Chuck decided to make his press of welded steel construction. Part way through the semester, as Professor Hall saw what Chuck was up to, he suggested that Chuck enter his design in the Lincoln Electric Arc Welding Contest (sponsored by the James F. Lincoln Arc Welding Foundation, in promotion of careers in welding).

Chuck considered the idea and decided that his best chance at winning was to have a good

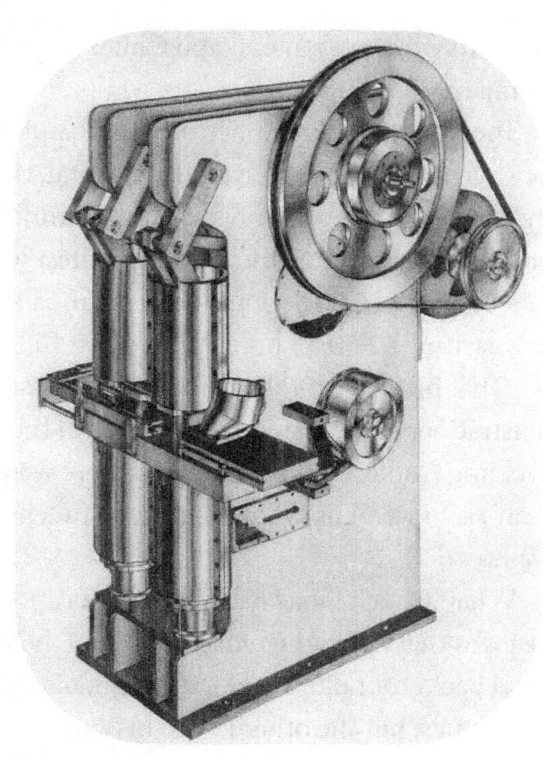

Chuck's design for the Lincoln Electric Arc Welding Contest with illustration by his wife, Ann.

---

v Herb Alvord and his wife, Lorraine, and Chuck and Ann became friends, and they remained in touch over the years. While Chuck and Ann are still connected with Lorraine as of the writing of this book, Herb passed away on May 8, 2018, three months short of his ninety-ninth birthday.

presentation. To him, that meant an excellent drawing of the finished press. And who better to help draw it than his artistic young wife?

"Ann is an incredible artist," says Chuck. "I asked Professor Hall what he thought of Ann and me entering the contest together. I could make a 3-D perspective drawing of the press, and Ann could do the job of shading it to look like a photograph." Professor Hall encouraged the young couple, and after several weeks of intense work together, they had an impressive entry. Ann Hutchins became the first woman to enter and win the Lincoln Electric Arc Welding Contest.

Chuck earned an A in each of his machine design courses and graduated from the University of Michigan in the spring of 1957 with a degree in mechanical engineering. Ann also graduated from Michigan that June, pregnant with their first child. The young couple moved into an Ann Arbor apartment, and Chuck got a job at Buhr Machine Tool Company.

The following September, as Chuck and Ann were bringing their newborn, Linda, home from the hospital, the phone was ringing. It was the University News Service calling to announce that Chuck and Ann had won first place in the welding contest with a $1,250 prize. The prize money was, as Chuck puts it, "like manna from heaven." At the time, he was earning $2.50 an hour in his job at Buhr.

"That picture won the prize!" Chuck insists. Ann's beautiful illustration was featured on the cover of the Lincoln Electric Arc Welding Contest publication, and there were articles about Chuck and Ann's achievement in the Michigan student paper and the *Ann Arbor News*.

When Chuck looks back on his education through his college years, he jokes that he graduated in 1957 BC ("before computers"). He had a good background in machine design, manufacturing, and automobile mechanics, but the other key to his career—computing—was yet to come. "Three words I didn't know when I graduated," says Chuck, "were *computer, binary,* and *entrepreneur.*"

2

Discovering Numerically Controlled Machine Tools

Buhr Machine Tool Company was an impressive outfit. Located on Green Street in Ann Arbor, along the railroad tracks just north of Michigan Stadium, the building took up nearly an entire block. Inside, the ceilings were five times as tall as a man, and walls of windows ran the length and breadth of the concrete room. Sunlight streamed into the cavernous space as Chuck passed some of the largest machines he'd ever seen. Men in aprons could hardly hear one another as the noisy equipment moved parts as big as car engines from one station to another on a transfer machine being built for Chrysler Corporation.

"They were installing a planer milling machine that was 105 feet long," remembers Chuck. "It had two 25-foot tables and four 75-horsepower milling heads. I mean, this was a major piece of machinery." It was so long—and had to be so precise—that it couldn't be leveled with a level because of the curvature of the earth; it had to be leveled optically.

Chuck wasn't seeking a job when he first visited Buhr during his last year at Michigan. By then, he already had job offers at Warner & Swasey in Cleveland, ExCello in Detroit, and the General Motors Tech Center in Warren. (Then, as now, U-M engineering grads were a hot commodity—even those with average grades.) Still looking for his best option, Chuck was curious about another machine tool company— Cross Company in Detroit—where Mike Zajac, the chief engineer at Buhr, had worked previously. Chuck requested a meeting with Zajac to ask some questions about Cross.

After the two men talked for a bit, Zajac asked, "Why aren't you interested in working for us?"

Buhr Machine Tool Co., established in 1925 by Joseph F. Buhr, made primarily high-speed, multi-spindle drilling machines for the automotive industry. The majority of the machine tools they built were transfer lines (a group of machines working together in a production sequence with precise positioning of the parts) and dial or trunnion machines that moved a stationary part through a rotary of spindles, either like a merry-go-round (dial) or like a Ferris wheel (trunnion).

Buhr was a strict, well run, profitable company. It had about fifty engineers on staff, though none in 1957 were college educated. Chuck's salary as a starting engineer at Buhr would be $2.50 an hour, half of what his other offers were at that time. "But there was something exciting about that little company in Ann Arbor," Chuck remembers. Zajac set up a meeting for Chuck with the executive VP of Buhr, Wilbur Gerchow, who confirmed that Buhr was interested in hiring him.

Chuck confirmed he was interested in working there, but if he was going to accept the low pay, he had a counteroffer. Chuck remembers, "It's hard to imagine how I approached the subject then, as I was a pretty shy, unsure guy, but I explained that I would accept his offer if he would do something for me." For the first six months, Chuck wanted to work one week each with the twenty-six most experienced, skilled men in the plant. After years of theoretical learning in college, he wanted a more practical understanding of machine tool manufacturing. Gerchow agreed, and in July 1957, Chuck arrived at Buhr as the first degree-holding engineer they'd ever hired.

For the rest of that year, Chuck experienced almost every aspect of building machine tools. "Forty hours, one-on-one, with twenty-six outstanding machinists and machine tool assembly people. For me, that was like a PhD degree!" Most of the machinists welcomed his interest in their work; some even let Chuck experience running their machines. One time, when he was cutting teeth on a gear using a gear hob, he put the change gears on backward, and the machine started milling off the gear teeth it had just cut. Chuck laughs and shrugs at the memory: "You show me a guy who never made a mistake, and I'll show you a guy who never did anything. But it did bring me up short."

Another week, he was watching a machinist, Hank Braunz, operate a very large, precision horizontal boring mill right outside the foreman's office. The rough steel weldment that was being processed by this machine was expensive, so when Hank encouraged Chuck to try his hand at the machine, the foreman walked out of his office and put a stop to it. "Chuck," he said, "it's okay to make a mistake on a few gear teeth, but that part is too big and too expensive for mistakes."

Though Chuck did learn a lot about machining, the most important benefit of his six months of exploration was building relationships with the machine tool operators. Though no one anticipated it at the time, those connections would be essential when Chuck started experimenting with numerical control.

But first he had to take on his duties as just another Buhr engineer. In early 1958, he settled down at a drafting board and started drawing. His new supervisor said, "Chuck, the only difference at Buhr between a degree-holding engineer and a non-degreed engineer is that you get a back on your drafting stool." For the next fourteen months, Chuck drew machine parts as he learned the intricate details of machine tools. "I was now drawing parts similar to those I had seen manufactured in the shop."

■　■　■

It was the fall of 1958 when Gerchow called Chuck into his office and said, "There's a seminar coming up at IBM in Endicott, New York. I'd like you to go." At that point, Chuck still had no experience with computers, but he flew to New York and attended the week-long seminar about the IBM 305 RAMAC (which stood for "Random Access Method of Accounting and Control").

This machine was an early computer designed to automate some elements of bookkeeping. The processing unit itself was about the size of a standing wardrobe—four feet wide, three feet deep, and six feet high. All closed up, it looked like a gray, metal closet, but inside were miles of electronic circuitry and a tall stack of rotating disks. The IBM 305 was the first computer with a rotating disk memory. Fifty platters, each two feet in diameter were stacked on top of each other in a spinning cylinder with read-write heads that moved in and out between the platters. Each platter held 100,000 characters, so fifty disks provided a total of 5 megabytes (MB) of memory. (Today, a single song on iTunes averages about 4 MB.) The IBM 305 rented for $3,200 per month and could be set up in a large office. The user worked at an adjoining desk with a keyboard, punch cards, and a printer. In short, this machine was a far cry from anything Chuck had worked with in the machine tool industry.

He remembers, "Almost every word of that seminar went right over my head. For a mechanical engineer with a major in machine design, it was all Greek to me. I felt lost!" When he got back home, he told Ann, "If my job depends on this, I'm in trouble."

The only part that did make sense was when Chuck learned that the rotating memory drum of a similar computer, the IBM 650, was made by Bryant Computer Products in Walled Lake, Michigan, which was a subsidiary of Bryant Chucking Grinder, a machine tool manufacturer in Springfield, Vermont. "BCG knew how to make precise spindles," says Chuck, "and if you want read-write heads to fly close to the surface of a rotating cylinder, you better have really good bearings." In other words, if nothing else, Chuck understood that computers required precision machining.[vi]

■　■　■

About six months later, on Friday, March 20, 1959, Wilbur Gerchow called Chuck into his office with another assignment: "There's a seminar at Purdue next week. The subject is numerical control. I don't know if we should build it, sell it, use it, or what, but you go and learn what it's all about." Gerchow's secretary had already made Chuck's reservations. Gerchow gave him a company credit card and the keys to a company car and said, "Just be in Lafayette, Indiana, by Sunday evening."

Chuck never found out why Gerchow asked him, of all the Buhr engineers, to attend these seminars. Perhaps it was because he had a college education, or because he had shown his dedication and curiosity when he did his six-month rotation in the shop. Or maybe it was something about the way Chuck's brain seemed to process information.

---

vi Chuck later learned that the University of Michigan had had an IBM 650 installed in the basement of the Rackham building while he was still a student there, but he never saw it or learned how it was being used.

Though neither man knew anything about computers at that time, Chuck would soon prove that he was able to program a computer. Could Gerchow have seen something in the twenty-five-year-old that pointed to this future? Who knows.

The 1959 seminar was officially billed as a conference on "manufacturing automation" organized by the School of Mechanical Engineering and the Department of Industrial Engineering at Purdue. It was the third such conference, following gatherings in 1956 and 1957, and it involved several presenters from IBM as well as corporate leaders and engineers from manufacturers like General Electric, Delco Remy, and Universal Controls, Inc.

It was Wally Brainard, the chief engineer of the Servo Machine Tool Division of Kearney & Trecker Corp., who was about to change Chuck's life. Like Buhr, Kearney & Trecker built milling machines. Brainard's presentation on Wednesday, March 25, 1959, was called "Numerical Control of Production Machines." The conference program described the session this way:

> *The most important development in production machinery in 80 years, numerical control is now off to a flying start. Systems in use number in the hundreds. What are the attributes of a good practical system? What advantages are possible for small and variety-lot production? Dollar savings?*

At the time of this presentation, there were fewer than 400 numerically controlled machines in the world. Most were in government-funded aerospace plants, and few engineers had seen one in action. Like his boss, Chuck had no idea what numerical control (NC) was, or what it was good for, but he was about to find out.

■ ■ ■

The invention of numerical control goes back to the late 1940s with the work of John Parsons, owner of a small manufacturing firm, Parsons Manufacturing Co., in Traverse City, Michigan. During World War II, Parsons and his father built landmines and bombs, and in 1944, John secured a contract to build rotor blades for helicopters. As he worked to improve the template design for the blades, Parsons and aeronautical engineer Frank Stulen discovered how to calculate airfoil coordinates—a complicated and time-consuming process—using an IBM 602A electromechanical calculator. (Don't let the word "calculator" suggest something handheld; this was a machine about the size of an upright piano.) When they fed these data—via punch cards—into a Swiss boring mill, they had created the world's first numerically controlled machine. Parsons called it the "Card-a-matic Milling Machine."

Starting in 1949, Parsons worked with researchers at MIT to advance the technology. In fact, it was a naming contest at MIT that resulted in the moniker "numerical control." They soon discovered that punch cards slowed down the process, so the cards were replaced with punched paper tape, and a paper-tape reader was installed on the milling machine. Although MIT began to compete with Parsons for relevant contracts and for the patent, Parsons and Stulen would eventually receive the first NC patent in 1958.

As with Parsons's helicopter blades, the early development of NC technology was driven by the aerospace industry. Advances in aerodynamic wings for fighter jets and passenger airplanes called for the machining of precise parts along the X, Y, and Z axes. Thus, funding for these new aircraft also funded NC research. It was a combination of the U.S. Air Force, other government agencies, and aerospace companies that supported MIT's work on APT, a high-level programming language developed in the late 1950s to generate instructions for numerically controlled machine tools. As the aerospace companies saw the potential for NC technology, various machine tool builders began to respond.

When Chuck saw the 1959 presentation at Purdue, Wally Brainard of Kearney & Trecker was talking about his company's new NC machine, the Milwaukee-Matic—a numerically controlled machining center that could mill, drill, tap, or bore, automatically moving the part along the X and Y axes or rotating it around the Z-axis to work on all four sides. Chuck immediately grasped the basic potential of NC, and he went back to Gerchow to say, "There's something here."

"Go out and become the expert," Gerchow told him.

Chuck spent much of 1959 traveling the country to visit NC users. "My objective was to learn as fast as I could, as much as I could." He visited Boeing and Rocketdyne (where he saw an early rocket engine and met Dick Stitt, who would later be his colleague) as well as machine tool builders that were incorporating NC into their production, like Burgmaster, Sundstrand, Giddings & Lewis, Kearney & Trecker, and Cincinnati Milling Machine (the precursor to Cincinnati Milacron).

Though the total number of NC machines in use was minimal, there were as many kinds of NC machines as there were machine tool companies, each experimenting with its own designs. As with any new technology, the number and variety of NC machine types would eventually consolidate into a few versatile approaches built by experts in the industry. But until then, Chuck could observe many different solutions to the same basic problem.

"For the first time, I was seeing machine tools operating without an operator turning the cranks," says Chuck. "The value of my six months in the Buhr shop became obvious, because I understood what I was seeing. And I was seeing something revolutionary." Chuck remembered back to those endless days, ten years earlier, when he made 1,125 light bulb sockets on the South Bend lathe in his basement. "I thought, *Now we could write a program and a machine will just keep popping those light bulb sockets out all on its own?* It was like a hard right turn."

3

Bringing Numerical Control to Ann Arbor

When Chuck and his colleagues at Buhr got their hands on their first NC machine, it was something of a disappointment. In early 1960, Pratt & Whitney Machine Tool advertised a numerically controlled point-to-point drilling machine, the Tape-O-Matic, for just $8,595. "The machine was so relatively inexpensive that you couldn't afford not to give it a try," says Chuck.

It even came with a money-back guarantee: "If by 90 days after purchase, this drill has not reduced your drilling costs, return it for refund, less transportation costs." Says Chuck, "That was unheard of in the machine tool industry."

When the Tape-O-Matic arrived at Buhr, Chuck was there to inspect it. What he found was basically an ordinary drill press head (hidden inside a modern welded steel enclosure) with power feed and a Morse taper spindle (the tool-holding device). Chuck was not impressed. In the previous ten months that he had been traveling the country to see NC machines at work, he had learned four secrets to successful NC:

- every step to manufacture a part was preplanned,

- the numerical dimensions were punched into a 1"-wide paper tape (the punched holes reminding him of the player piano scroll he knew from childhood),

- the tooling necessary was specified in advance and available in a tool rack dedicated to the specific job, and

- the work holding method—how the part would be secured to the machine during the work process—had been determined and was available at the NC machine before the parts arrived.

Says Chuck, "The Tape-O-Matic, as delivered, satisfied none of these requirements."

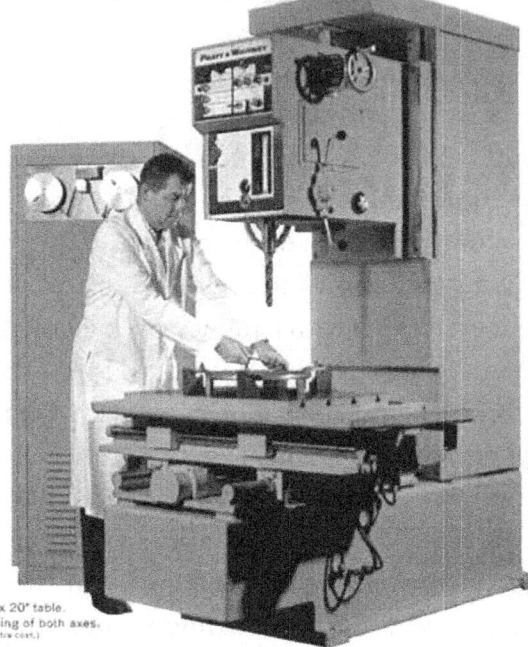

HOT TECH COLD STEEL

But Chuck had some ideas to improve it. He designed, and Buhr manufactured, a custom Universal Kwik-Switch spindle to replace the unwieldy Morse taper spindle. Then Buhr bought Universal Kwik-Switch tooling to use in the newly installed spindle. That solved the problem of having all the required tools for a given job in a dedicated tool rack at the machine tool when the job arrived.

To ensure an efficient and accurate work holding method, Chuck asked the shop to build two fixed vise jaws and two movable vise jaws that could be quickly adjusted using air cylinders. "That way," says Chuck, "all you had to do to clamp the part to the machine was throw the air valve. Then we acquired and installed two precision zero position indicators so the machine could be positioned quickly to the zero intersection of the fixed vise jaws. Only then did I feel we were ready to give NC a fair try."

The test job on the Tape-O-Matic would be simple: drill three equally spaced holes on a round part held in a three-jaw lathe chuck mounted on a 10"-square plate placed in the vise jaws. No rotation was involved. All the machine had to do was get the tool in the right location and drill three holes. Says Chuck, "That was something we could program."

The process engineers numbered the holes for the program, and Chuck's secretary entered the instructions into a Friden Flexowriter—a heavy-duty typewriter that punched out a 1"-wide paper tape. (The simple programs for the Tape-O-Matic resulted in punched tapes that were one to four feet long.) The tape was then fed into the upgraded Tape-O-Matic, and—voila!—the machine began drilling holes into the round part, one after another.

Says Chuck, "It wasn't exactly what Pratt & Whitney had in mind, but I do believe the machine paid for itself in the first ninety days of operation."

In his new role as Buhr's NC expert, Chuck was seeing that his initial six-month apprenticeship in the machine shop was paying off, not only because his relationships with the guys in the shop made it easier to ask for their help with these NC experiments, but also because he had to think like a machinist. "When you write a program to run an NC machine tool, you *are* the machinist," says Chuck. "And when I didn't understand something, I had twenty-six guys in the shop I could ask."

. . .

The purchase of Buhr's second NC machine was announced while Chuck was on the road. When he returned, he was shocked to learn that a numerically controlled DeVlieg JigMil would arrive in November. Chuck thought, *Why is Buhr buying another NC machine without asking my advice? I thought I was supposed to be Buhr's NC expert.*

It turned out that Charlie DeVlieg and Joe Buhr were good friends, and they were having dinner one night when Charlie sold Joe on the idea that Buhr should have the very first numerically controlled DeVlieg JigMil—Serial Number One. The symbolism of such a commitment did make some sense, as Buhr Machine Tool Co. had been an early customer of DeVlieg Machine Co. of Ferndale, Michigan.

Founded in 1939 by Charles DeVlieg and his son, DeVlieg Machine Co. made its name with the invention of the DeVlieg JigMil—a precision boring and milling machine. As the story goes, Charlie had created a predecessor machine tool that was capable of producing aircraft-engine supercharger blades fifteen times faster than other machines. During World War II, he had so many orders for this machine that he couldn't wait for the delivery of two new horizontal borers, so he set about to build his own. The result was the DeVlieg JigMil. As DeVlieg's promotional materials explained, "The advanced structural design of the JigMil provides a fixed relationship between the horizontal and vertical slides which ensures precise accuracy in the full range of the machine." Customers used the JigMil to make a variety of production parts like gear cases, aeronautical parts, printing presses, and paper and textile machinery.[2]

Buhr had long used several DeVlieg JigMils in its shop, so Joe Buhr was interested to see how DeVlieg's numerically controlled version would operate. Joe agreed to pay $25,000 for the first NC JigMil. Alas, it was a work in progress. Demonstration of the machine was scheduled for the 1960 International Machine Tool Show in Chicago, but it wasn't running by the time it was supposed to be shipped to Chicago that June. Instead, it arrived at Buhr in November of 1960.

HOT TECH COLD STEEL

As Chuck remembers, "It became my job to make it run." He quickly realized that the NC DeVlieg JigMil was another good start but wasn't totally numerically controlled either; the only real advancement was that the XY coordinates were entered automatically from the punch tape. The Z-axis—the spindle that moved in and out of the part—still had to be controlled by the operator or by mechanical "trip dogs" that tripped a switch. Says Chuck, "You couldn't put the machine in automatic cycle, walk away, and come back to a finished part. While it was a good learning experience, I'm glad the decision to buy it wasn't mine." [vii]

The DeVlieg machine helped Chuck explain to his superiors at Buhr that a totally numerically controlled 3-axis machine was the key to a more productive future.

■   ■   ■

In the early spring of 1961, after more than two years of research, Chuck proposed to Wilbur Gerchow that Buhr should buy a Sundstrand Model 21 vertical spindle, twin column, rail machine with a full 3-axis continuous path control and an automatic tool changer for up to twenty tools. With an 84" X-axis and a 40" Y-axis, the Model 21 was big enough to handle the cases for fixed spindle drill heads, a key product made by Buhr. In addition, it would be one of the very first NC machines ever made with a transistorized controller.

Gerchow liked the idea, but the price tag was another matter: $150,000 (about $1.3 million in 2020 dollars). That was six times the price of the NC DeVlieg JigMil. "For a purchase like that," said Gerchow, "you'll have to make a presentation to the board of directors."

Buhr's board included two men in leadership at the University of Michigan—Roscoe Bonisteel (a regent) and Wilbur Pierpont (vice president of finance). (If these names sound familiar, it's because Bonisteel Boulevard is now a main road through U-M's North Campus and passes right by Pierpont Commons, the North Campus student

vii Though that first NC DeVlieg machine was rather lacking, the company would go on to build highly sophisticated numerically controlled machines, especially during the tenure of Richard Jerue as DeVlieg's vice president of engineering (Jerue was just sixteen years old when he started working for Charlie DeVlieg). Chuck was particularly impressed with a DeVlieg NC machine built for the oil industry that could mill large, steeply tapered internal threads precisely and uniformly every time.

union). In 1961, neither man was an expert on machine tools, but they must have been impressed by Chuck's sales pitch.

"I tried to explain how automation was going to raise the productivity of the shop," Chuck remembers. "I told them why I thought the Sundstrand machine was the one for us. I know they didn't understand all I was saying, and I only understood maybe half of what I was saying. But I must have succeeded, because Buhr placed the order." It was one of the first Sundstrand machines ever sold to a non-aerospace company.

Chuck later learned that both Michigan men voted for the purchase. He believes his engineering degree from the University of Michigan probably helped seal the deal.

The Model 21 would be built at Sundstrand Corporation's machine tool manufacturing facility in Belvedere, Illinois. Buhr wouldn't receive it until September 1962. "For more than a year, we were like new parents, preparing for its delivery," Chuck remembers.

■　■　■

His first challenge was to figure out how to program the new machine. He was trying to do so manually when Frank Westervelt heard what Chuck was up to. Westervelt had been a teaching assistant in the College of Engineering when Chuck was a student. In 1961, Frank was just completing his PhD in mechanical and electrical engineering; he would become assistant director of the U-M Computing Center. As Chuck told Frank about his need to program a new NC machine, Frank said, "You need a computer."

Chuck laughed. "Why do I need a computer, Frank? I've got a Friden Flexowriter for punching tape, a Friden Square Root calculator, and Hans Hof's seven place trig tables."

It didn't take Chuck long to realize Frank was right. "The whole programming activity was painful!" Chuck remembers. "But then I remembered the IBM 305 RAMAC!" That 1958 seminar about a bookkeeping computer hadn't made much sense back then, but now that Chuck needed to generate tapes for NC machine tools, the power of computing was suddenly apparent.

　　　　　　　　　　　HOT TECH COLD STEEL

Chuck learned that Sundstrand had already developed a computer program called SPLIT—Sundstrand Programming Language Internally Translated—for use in programming Sundstrand NC machines. It was authored by a Sundstrand engineer, Harold "Hal" Baeverstad, who had earned his engineering degree at Northwestern University after winning the bronze star for bravery during World War II.

The earliest programs for machine tools were MIT's APT (Automatically Programmed Tool), written in the late 1950s, and IBM's ADAPT, developed in the early 1960s. Those programs ran on large mainframe computers that only very big operations, like aerospace companies and the military, could afford. The programs also used formal mathematical descriptions of geometry that required a programmer to have strong mathematical skills. As a result, APT and ADAPT did not make much sense to the typical shop machinist who understood machining practices and trigonometry but did not typically have a strong math background.

In contrast, SPLIT used abbreviated machine shop words to describe the motions and auxiliary functions of the NC tools, such as spindle speed, feed rate, coolant, and spindle direction. The coded language was self-explanatory, with instructions like ATCHG (Automatic Tool CHanGe), IPM (inches per minute cutting feed rate), and RPMRH (revolutions per minute right hand—to start the spindle going in a clockwise direction).

Chuck says, "Hal Baeverstad was an outstanding programmer." Baeverstad created SPLIT on the IBM 650 and later rewrote it to work on the IBM 1620, as these early computers advanced. As Sundstrand was building Buhr's machine, Chuck was able to acquire a copy of SPLIT from Sundstrand (and, later, the source code as well). "Having a live example of a real computer program was a serious learning aid," he says.

However, Chuck had no experience working on the IBM 1620—or on any computer. He called IBM for help and soon met Ed Downing, an IBM customer-service engineer working out of Detroit. There was an IBM 1620 available for use at the IBM Education Center in Detroit, but the only time Chuck could reserve it was in the middle of the night. Ed Downing would meet him there.

"We struggled together," Chuck remembers. (The two men remain friends to this day.) "Hardly anyone knew how to program a computer then, even most IBM employees." Chuck used the manual for the IBM 1620 Symbolic Programming System (SPS) to slowly unravel the mystery of this early programming language. "SPS was not a higher-level language, but I didn't even know that term back then."

The IBM 1620 was a decimal digit machine with 20,000 six-bit characters of memory. (Note that 20,000 is a decimal number, rather than the binary number 20,480.) Data would take up some of that space, but even if the full memory was available for instructions, most instructions required twelve characters, so the memory would be used up with only 1,666 instructions.

Says Chuck, "The 1620 was jokingly called the CADET computer, an acronym for Can't Add, Doesn't Even Try. It had two look-up tables in low memory, one each for addition and multiplication. The hardware looked up the answer, one digit at a time, including the use of a 'carry bit,' just like we learned in elementary school."

The 1620 ran on punch cards, created on a key punch. But machine tools ran on punched paper tape, so Chuck also needed an IBM card-to-tape converter. When that machine got out of adjustment, it would throw the cards on the floor, so he also needed a card sorter to keep the cards in order. There was also a separate printer to print reports, the IBM 407. In other words, making software that would run on the IBM 1620 with output that would run an NC machine tool was a real challenge. And that was before Chuck tackled the task of writing a program that would actually work.

Chuck asked Ed, "How do you ever learn how to program?"

Ed replied, "Pick something you know how to do and teach the computer how to do it, one step at a time."

With this advice, Chuck conceived a program to machine the cases for multiple spindle drill heads. He knew this was the kind of job that the machinists at Buhr would want the Sundstrand 21 to do. "There were six different size drill head castings," Chuck remembers. "Each had a length and width, two hole locations for the dowels to align the front and back of the case, and the required screw locations to hold the two pieces together. I thought it would be easy to put that information into a table. Then, when the operator picked a casting size, all the machining would just happen. In addition, each shaft in the drill head had an XY-coordinate, a ball bearing size, and a number of bearings deep in the hole. With that information included, the program could go ahead and finish machining the rest of the casting without further programming. The whole idea was to design the instructions up front to minimize the number of manual setups required to finish the part."

When he explained to Ed Downing what he had in mind, Ed said, "For Christ's sake, Chuck, can't you pick something simple?" But Chuck knew this was a problem he needed to solve to make NC a real success at Buhr.

Reserving computer time late nights in Detroit was not practical on a continuing basis (especially now that Chuck and Ann had three children under age five). So Buhr arranged to share the rental costs for an IBM 1620 (about $3,000 per month) with another Ann Arbor company, Conductron Corporation, led by Kip Siegel, a U-M physics professor. Conductron would become a pioneer in holography research; at that time, it was developing radar for the military. The IBM 1620 was installed on the second floor of the sprawling West Huron Street headquarters of American Broach & Machine Company, which had just been purchased by Sundstrand. Chuck shared his time on the machine with Conductron's engineers, including Mike Levine (a designer of an early microprocessor, who would invent VCR programming, the programmable thermostat, and other patented technologies). When Conductron built a new headquarters on Green Road in north Ann Arbor, the IBM 1620 was moved to that location.

With more time on the 1620, Chuck began to make progress with his machine tool program. When he encountered a problem, he often found that sleeping on it would reveal a solution. (His daughter Linda claims he started sleeping three times a day so he could solve more problems.) By the end of the summer of 1961, after completing the simplest part of the operation, Chuck named his program "Buhr Automatic Tool Selection And Split Statements"—or BATS-ASS for short.

A few of the other engineers at Buhr, including Jack Clausnitzer, Dave Bernhardt, and Dave Stormont, showed interest in Chuck's work, and he walked them through the SPLIT instructions and the BATS-ASS program. Bernhardt and Stormont would eventually start their own business, NuCon, in Livonia, Michigan, using Sundstrand 5-axis Omnimills. Jack Clausnitzer and his wife, Heidi, would eventually open their own NC machine shop, Brighton NC, in Brighton, Michigan; it is led successfully today by their three children.

■ ■ ■

In addition to programming, Buhr's preparation to receive the Sundstrand Model 21—the first NC machine controlled by transistors—included training in how to maintain the controller (the part of the machine that read the punched paper tape and told the machine what to do). Built by TRW, Inc., in Michigan City, Indiana, the transistorized controller contained printed circuit boards about 10" x 20" with fifteen to twenty-five discrete transistors per board. Each transistor was about the size of a pencil eraser sitting on three wires. (Compare that to the 7 nanometer chip in a 2019 Apple iPhone 11 Pro that contains 8.5 *billion* transistors.)

In the spring of 1962, Chuck and Bob Harte (an electrical engineer at Buhr) attended a TRW controller maintenance class. All Chuck knew about transistors was what he'd learned in the required U-M electrical engineering course EE7 that he'd taken in 1956. Bob, a former journeyman electrician with Chrysler, was no transistor expert either, but he had designed the relay logic circuitry for the massive transfer lines that Buhr manufactured. The transfer line controller had miles of wires connecting the relays. In other words, Bob Harte wasn't daunted by complex systems, nor were the guys in the shop who followed Harte's ladder diagrams to build those controllers.

At the TRW controller class, Bob and Chuck received maintenance documentation for the transistorized controller that contained logic diagrams and wiring charts, but it wasn't presented in the familiar ladder diagrams that Harte's guys were used to working with. On the way home, Bob said to Chuck, "Our electricians will never be able to maintain that controller without better documentation. For me to understand it and for the guys in the shop to understand it, I suggest we go through the transistor logic and regenerate it all as ladder diagrams. That way, we can better understand exactly how it works." The task of converting the maintenance documentation to ladder diagrams took Bob two months, but it was worth it. Not only did he create useful documentation for his team, but in the process, he found a wiring error that TRW was able to fix before the machine was delivered.

Bob also made several little flip-flops that he could attach anywhere in the controller to see where a stray signal might be coming from. (Compare that to today's computer controlling software, hidden behind a user interface, that allows the user to query when any kind of failure occurs.)

■　■　■

Chuck was standing outside the Buhr facility on Saturday, September 1, 1962—Labor Day weekend and his sixth wedding anniversary—as a semi truck pulled into the parking lot to deliver the Sundstrand Model 21. It had been eighteen months since Buhr placed the order. It would take several more months to install, level, and get the machine running. But when they did, Chuck's programming worked. The sequence of steps contained in BATS-ASS generated the necessary SPLIT statements to run the Sundstrand machine, which completely machined both halves of a fixed spindle drill head without any human intervention.

The machine ran well for several months through the winter of 1962–1963, but as spring arrived, the temperature in the shop rose, and so did the transistor failures. An air conditioning unit on top of the transistor cabinet was supposed to regulate the temperature, but it was only circulating cool air at the top of the cabinet. Sometimes a low-tech solution is all that's needed: a cardboard baffle inserted down the center of the cabinet, held in place with masking tape, did the trick.

The machine was soon in full production, running twenty-four hours a day, seven days a week, with three shifts. Operators got paid for eating their lunch while the machine continued to run. Each operator got every third weekend off, with the other two operators working twelve-hour shifts. The paychecks were huge, the hours exhausting. But the production was impressive.

Needed less frequently in the shop, Chuck had the idea to use his newfound programming skills on another project: a cost accounting program for Buhr's vice president of finance, John Hamilton. Using the punch cards that the machinists had already created when they clocked on and off from making a particular part for a machine tool, Chuck figured he could write a program for the IBM 1620 to summarize how much it had cost to build a given machine tool. He remembers, "That program was relatively easy."

■　■　■

The success of the Sundstrand Model 21 soon led to discussions about acquiring another Sundstrand machine. The engineers were in agreement that the next model should be a horizontal spindle machine.

As the team developed the specs for this second machine, Gerchow questioned the value of the automatic tool changer (ATC). He said the operator could do that job as quickly as the ATC. Chuck disagreed, and he had data to prove it. "As it happened, when we were installing the first machine, I had the foresight to ask our electricians to install two running hour meters to accumulate 'total on time' and 'cycle on time' and a Veeder-Root counter to count the automatic tool changes."

When Gerchow expressed his doubts, Chuck asked permission to leave the meeting for a moment, rushed to the shop, gathered up three numbers, and ran some quick calculations. Upon returning to the meeting, he announced, "The ATC changed the tool every two-and-a-fraction minutes of cycle-on time since we began using the machine. There's no way our operators would have kept up that kind of average." Chuck agreed that an operator might change the tool faster at any instant if he was always there when the job needed doing, and if he always had the correct next tool in hand. But, thanks to the ATC, he was often doing other things, like bringing coffee to the other operators in that bay of the shop, while his machine kept going in his absence. Chuck added, "The ATC always gets the right tool from the right place, every time." Everyone agreed to keep the ATC.

With the Buhr board of directors in agreement, Chuck approached Sundstrand about building a second machine. The only difficulty was that Sundstrand didn't build a machine with the specs Buhr needed. Chuck remembers, "I became a salesman, trying to convince them to build a special machine for us."

Sundstrand did build a series of "modular" bed mills, and they also built a horizontal spindle machining center, called the "OM-2," but it could only handle parts that were maximum 2' x 2' x 2'. It was too small to meet Buhr's requirements. Chuck explained to Sundstrand, "What we want is a modular horizontal spindle machining center that would use the bed and table from the Model 21—with its seven feet of X travel— with the column, spindle, and tool changer from the OM-2. Oh, and make the column three feet taller." The result was Sundstrand's first "wide table" OM-2 #17 with 7' x 5' of XY-travel and 2' of Z-travel and a forty-tool automatic tool changer. It became a highly successful machine for both companies.

The custom machine was quite an investment for Buhr, costing about $350,000. But, like the Model 21, it was soon paying for itself in vastly increased production. Word got out in industry circles about this massive NC machine at Buhr, and engineers came from throughout southeast Michigan to see it. Chuck found himself giving tours to men from General Motors, Lincoln Park Boring, and other companies. After six years of experiential learning, he had, in fact, become an NC expert.

■ ■ ■

One of the first jobs Buhr pursued with the OM-2 was a suggestion from Chuck. "It was a real advantage to have a machine design background," he says. "Buhr built some palletized transfer machines, and I suggested we build the same pallet clamping mechanism into a pallet changer for the new wide table OM-2." They put the strategy to work on a contract with Oldsmobile to build a palletized transfer machine to make steering knuckles. Buhr machined the base, mounted the steel pallet base plate with four locator bushings, and then did all the machining relative to the base plate, all while handling the fixture on the automatic pallet changer.

Chuck remembers, "We saw a key benefit of the pallet changer when, in assembling and testing the transfer machine on the assembly floor, the assembly people found an interference between the fixture and one of the boring heads of the transfer machine. Because we had the pallet changer on the NC machine, as fast as they could transport a pallet fixture from the assembly bay back to the OM-2, we could cut the clearance for the head and send it back to the assembly floor." Without the automatic pallet changer, that fix would have taken days, not just hours.

The second big job was machining a mechanical drive transmission for the slide units of a transfer machine. These were done in batches of thirty. Prior to NC, it took ninety days to move a batch through all the manufacturing operations. Once the OM-2 was programmed and tooled, and using the pallet changer, each batch of thirty was completely finished in five days.

The real win in using the pallet changer was being able to set up the next piece to be machined on a pallet while the OM-2 was machining the current piece. Thus, the idle time between parts became only the time it took to unload one pallet and load the next, just a fraction of a minute. Though pallet changers are common on today's computer numerically controlled (CNC) machine tools, they were a new concept in the early 1960s. But because Buhr already built palletized transfer machines, it was just a logical extension.

Given the successes of the first two Sundstrand NC machines, Buhr soon bought a second one of each—another Model 21 and another wide table OM-2. Having invested well over one million 1965 dollars into the new technology, by the end of 1965, the Buhr Machine Tool Co. was running four NC machines round the clock. In addition, the Pratt & Whitney Tape-O-Matic continued to pump out the small, simple jobs that required just a few drilled, tapped, or counterbored holes. The increased efficiencies obviously allowed Buhr to accept and complete a lot more jobs.

Looking back on that time, Chuck is reticent to call attention to his contributions to the company's success. He prefers to focus on the learning experiences he was afforded through his assignments. And just like his fifteen-year-old self was excited to figure out how to efficiently punch holes in sheet metal, thirty-year-old Chuck found great satisfaction in the problem-solving aspects of his job. "I enjoyed every minute of the long hours I put in making NC a reality at Buhr. While it wasn't always easy, it was always fun and challenging. I couldn't believe a person could have a job that he enjoyed so much."

4

The Programmer

Even as NC machining was taking off at Buhr, programming was "still a bear" for Chuck and his small team of engineers. They continued to work with the IBM 1620 housed at Conductron. Every new job required a new part program and a new punched paper tape. All testing of the accuracy of the program occurred at the machine tool itself—and all revisions required another trip to Conductron.

But Chuck was about to get some help, thanks to a phone call from Frank Westervelt. "It was one of the most significant events in my professional career," Chuck remembers. Frank told Chuck that he had a student at U-M, Bruce E. Nourse, who "stands tall among his peers" and was looking for a job. Bruce was twenty-three years old and had just completed his engineering degree. In the spring of 1965, Chuck hired Bruce to work at Buhr.

■　■　■

Westervelt had good reason to be impressed by Bruce, who had basically been closing in on the intersection of engineering design and computer science since he was a boy. Born during World War II, Bruce lived with his mother and grandparents in Florida while his father was away in Chicago working on the top-secret Manhattan Project. "My dad was a tool-and-die man," says Bruce, "and he helped design the bomb-dropping mechanism that dropped the bombs on Nagasaki and Hiroshima." Bruce only learned about his father's involvement in the atomic bomb program decades later. "He never, ever talked about it."

After the war, the Nourse family moved into the affordable barrack housing near the former Willow Run bomber plant in Ypsilanti, Michigan, so Bruce's dad could study aeronautical engineering at U-M. While a full-time honors student with a heavy course load,

Mr. Nourse also directed U-M's wind tunnel lab, conducting experiments in airplane and automobile design.

Bruce's mom, who had a college degree in psychology, became a secretary and bookkeeper for the Kaiser-Frazer automobile company when it was building cars at Willow Run. Young Bruce was fascinated by his mother's IBM proportional-spacing typewriter and by her ability to type a document while carrying on an unrelated conversation.

Bruce's playmates at Willow Run were a mix of kids from U-M families and auto-factory families. When he was ten, the Nourses moved to a house in Ann Arbor, where Bruce lived until he graduated from college.

During those years, Bruce's father ran his own engineering consulting firm out of a home office. Government contracts included a job to design rubber, disk-shaped fuel containers that could be thrown from airplanes, land safely on the ground, and be picked up by soldiers to refill Jeeps and tanks on the battlefield. Bruce remembers, "I went down to Dayton, Ohio, with my dad to watch them test these things. It was all my dad's design." Other contracts were for camshaft designs for engines of all sizes, from a chainsaw engine to a Cummins diesel engine. Says Bruce, "The camshaft in the Cummins engine was twenty-four feet long with valve heads the size of dinner plates. Those engines were designed to run for years and years without ever shutting down; you could replace the filters or the oil without shutting it down."

Bruce also remembers his dad designing the engine camshaft for a six-cylinder Jeep engine built by Willys Motor Company in Toledo, Ohio. "It only had six lobes on the cam," says Bruce. "The same lobe that was used to run the intake valve was also used to run the exhaust valve—one cam lobe ran both sides of the engine. That's very, very unusual."

■ ■ ■

Bruce started working for his dad as a teenager, doing math calculations and drafting at a fifteen-foot-long drafting table in the basement. "But the thing that sticks in my mind is the computer programming," says Bruce, whose first exposure to Fortran came in 1958, just a year after the programming language was first published (and the same year

Chuck attended his first computing seminar at IBM). "We also used a language called MAD," Bruce remembers, "which stood for 'Michigan Algorithm Decoder.'"

One of his dad's computer programs used 2,000 punch cards. Bruce was sometimes asked to make changes to this program at the U-M computing center in the basement of the engineering building. "One time," Bruce remembers, "I dropped the box. Luckily, it had been sequence ordered, so the cards were numbered. I could put the pile through the card sorter to get them back in order. But I had just punched about one hundred cards that weren't yet sequenced, so I had to figure out where they went." It was at the computing center where Bruce saw his first computer—an IBM 650 (the model used by Hal Baeverstad to program SPLIT)—and where Bruce worked on an IBM 709 and an IBM 709T, a transistorized model (also called the IBM 7090).

Bruce continued to work for his dad and live with his parents while studying engineering at U-M. He enrolled in a new program called "science engineering" that allowed him to take courses across the entire curriculum—mechanical engineering, thermodynamics, nuclear engineering, aeronautics, and so on—before settling into his choice of electrical engineering and engineering mechanics. "It was challenging," says Bruce. "I was taking fourth-year electrical engineering classes with people who had studied nothing but electrical engineering the previous three years." At that time, U-M was rumored to fail one-third of all freshman engineering students, and Bruce says that fully half of the science engineering students didn't make it. "I was not the world's greatest student," he says, "but I did well enough to graduate."

He took only one basic computer programming class. "I'm ashamed to say I did very poorly in that class." The key assignment was to write a program to figure out what set of coins would add up to a particular amount of money. Bruce remembers, "Most people start subtracting the biggest coins. It's very simple. But I didn't do it that way. I came up with what I thought was a much cleverer way with much less computer programming. My program worked, and it got the right answer, but I think the teaching assistant didn't understand what I had done. I think I actually confused him."

Bruce particularly remembers a course with Bernard Galler, a mathematician who taught the first programming course at U-M in 1956. Galler also wrote the programming language and compiler MAD (Michigan Algorithm Decoder) that rivaled Fortran at the time, and he would soon found the U-M computer sciences department. "Bernie Galler taught us about things coming in the future that we see today. He was right on the edge of knowing about neural networks and artificial intelligence, way back then. It was very interesting to me."

■　■　■

After graduating in December 1964, Bruce kept working for his father until one day his dad said, "Okay, Bruce. It's time for you to move on." That's when Bruce got an interview at Buhr. He remembers, "The first time I met Chuck Hutchins, he was out in the shop, with one foot up on the OM-2 and a clipboard on his knee, taking notes. He was wearing a suit and bow tie and leather shoes, go-to-church clothes, the kind of clothes nobody in their right mind would wear in a machine shop."

Thanks in part to Frank Westervelt's recommendation, Chuck hired Bruce on the spot as a part programmer. Says Bruce, "I did that for a little while, but I didn't really like it." When he expressed his dissatisfaction, Chuck said, "Well, I've got this other computer programming job, would you like to try it?" Chuck took Bruce to see the IBM 1620 at Conductron.

As soon as Bruce figured out how to write programs on the IBM 1620, he was hooked. "That was my thing," he remembers. After working all day at Buhr, he'd arrive at Conductron at 5:00 p.m. and work on the 1620 until eleven at night, every night. By then, Chuck had obtained the source code for SPLIT (Sundstrand's programming language for its NC machines), and Bruce figured out how to rework the code for more functionality. "I read that SPLIT program from one end to the other, and it was a great learning tool, because Hal Baeverstad was a really good programmer. He understood how to get the most out of a computer in the minimum space. When Chuck would ask me, 'Could we make the OM-2 do this?' I would figure it out. I added a few functions that were not in the original SPLIT program."

Bruce Nourse, 1974. The computer behind him is a Texas Instruments TI-980.
Photo courtesy of Bruce Nourse.

Chuck believes Bruce may have written the first-ever computer-aided design program during his time at Buhr. It required the use of a Calcomp plotter, which drew a graphic representation of a computer program on a three-foot wide roll of paper (similar to the way a pen moves across a seismograph). Bruce took the ladder diagrams that Bob Harte's team used when building a transfer machine controller, and he wrote a program that told the plotter how to draw the ladder diagrams. Anytime revisions were made to the miles of wiring on the controller's relay boards, Bruce's program (called the BEAST, for Buhr Electrical And Schematic Translator) could draw a new ladder diagram, complete with a new set of numbers for the wires (to keep track of how the wires were connected). Bruce had to write these plotter routines from scratch because none existed before that.

He also used the computer for entertainment purposes. Bruce remembers a funny example that was typical of early computerized tomfoolery:

After I'd been successful with the 1620 for quite a while, Buhr bought their own 1620, so we had one in-house. At one point, I wrote a program that made it look like the computer was answering your questions. I could type any yes or no question, and the computer would give a yes, no, or maybe answer. I programmed it with lots of yes-type replies, like "yeah" and "sure." And I had a lot of versions of no, like "nah" and "nope." And then I had a bunch of maybe answers.

My program was designed to give yes answers if I didn't enter any spaces after the question mark. But if I typed one space after the question mark, it gave me a no answer, and if I typed two spaces after the question mark, it gave me a maybe answer. Of course, anybody watching me type would not see the spaces, so I could make the computer answer any way I wanted.

Word got around the company, and people would come in and ask me to ask the computer questions like "Is it going to rain on Saturday?" If I'd seen the weather report, I'd use the space bar to answer yes or no. If I didn't know the answer to their questions, I could always type two spaces and get a maybe.

That program was a lot of fun until one night I was working late, and a janitor came in, and he wanted to ask the computer whether a particular horse was going to win an upcoming horse race. I was sweating bullets. You never saw so many maybe answers! We never could get that damn computer to give a definite yes or no on the horses he was interested in.

Antics aside, Chuck was stunned by Bruce's programming capabilities and was beginning to see what was possible with a truly gifted computer programmer. Today, more than fifty years since Bruce and Chuck first met, the two men are as close as ever. Bruce says that Chuck is "probably the most important person in my whole life in terms of where I've gone and what I've done." Chuck calls Bruce the "Michelangelo of software engineers."

They would share a lifetime of computing innovations. However, after working together at Buhr for just a few months, their collaboration took a brief hiatus.

Chuck cried on the day he walked into Wilbur Gerchow's office to resign from Buhr Machine Tool Co. It was just before Christmas 1965, and Chuck had been called home to help with the family business, the Pontiac Varnish Company. He had been serving on the small company's board of directors for about three years, trying to help his uncle Paul (the husband of Hutch's older sister Janice) keep the paint manufacturing business going.

But Paul was getting older, and the company hadn't kept up with the times. Eventually the board convinced Chuck that he should come in as executive vice president. Though Chuck knew nothing about making paint, none of the other grandsons of C.H. Hutchins had the potential to run the company. "I felt obligated to give it a try," Chuck remembers.

At Buhr, Gerchow told Chuck, "You're the only person who ever came in here to resign who had tears in his eyes."

At the start of 1966, Chuck reported for work at the Pontiac Varnish Company and immediately set about to make some changes. He hired a new technical director, enhanced the laboratory facilities, and improved some methods for formulating paint, especially for color matching. "After sixteen months, I'd learned a fair piece," says Chuck, "but I wasn't having any fun. Making paint wasn't nearly as much fun as working with computers and machine tools. So, I convinced the board to sell the company."

By then, it was early 1967, Chuck was back in Ann Arbor, and he was ready for a new challenge. "I asked myself, 'What are you going to do next?' and I came to the conclusion that if numerical control was ever going to be successful, other businesses were going to need help with the NC programming. I thought maybe I could take what I had learned at Buhr and help others avoid the NC part programming problems I had experienced."

Part II
The Start Up

5

Computer Timesharing

Chuck Hutchins was walking in downtown Ann Arbor in early May 1967 when he ran into Bob Guise on the sidewalk. Chuck was newly unemployed after leaving the Pontiac Varnish Company. Bob was CEO of the Ann Arbor company Comshare (earlier spellings were "Com-Share" and "ComShare"), which had just celebrated its first anniversary.

Chuck and Bob had first met when Chuck was still at Buhr. Bob, a 1954 graduate of the U-M College of Engineering, had been impressed by Chuck's knowledge of numerical control and by the enthusiasm Chuck exuded every time they met. This day was no exception. Bob remembers, "I had slipped out of the office for a quick lunch and to run an errand for my wife downtown. Near the corner of Liberty and Division Streets, I literally ran into Chuck. A brief apology and exchange of pleasantries placed us at lunch."

Chuck told Bob that he was leaving his family's paint company and that he couldn't stop thinking about how to make numerical control more useful to the metalworking industry. He took a chance and said to Bob, "I have an idea for programming NC machine tools using computer timesharing. Do you think Comshare could pay me just enough to feed my wife and three kids for a year and let me try it?"

It was an intriguing proposition with the potential to create a valuable customer base for Comshare's services. Bob Guise leaned over the table toward Chuck and said, "Tell me more."

■　■　■

Computer timesharing put a new spin on the progression of numerical control technology in the machine shop. And once again, the University of Michigan was foundational to that progress, as the founders of Comshare—Bob Guise and Richard "Rick" Crandall—had met at the U-M Computing Center. Their combined backgrounds in engineering, computer science, management, and finance fed a strong entrepreneurial mindset. They started their venture in 1966, and by early 1967, they had received a $1 million investment from a member of the Weyerhauser lumber family.

In today's connected world where everyone seems to have a phone, tablet, and laptop with continuous access to the web, it's hard to recreate the time when computers were rare and didn't know how to talk to each other. Comshare came about in response to the discovery of how computers could be connected across long distances.

The earliest research on computer timesharing came out of MIT and Dartmouth College, research that later allowed GE to offer computer timesharing for NC machine tools. But soon a pioneering group of researchers at the University of California–Berkeley (UC–Berkeley, or simply "Berkeley") took up the computer timesharing challenge. This effort started in 1964 and was known as Project Genie. It was funded by the Advanced Research Projects Agency (ARPA)—the U.S. Defense Department program that funded the development of the internet—and it involved some key individuals in the history of computing.

One of the professors who brought Project Genie to UC–Berkeley was Dave Evans who would later advance computer graphics through his company Evans & Sutherland. Others on the project were staff member Melvin Pirtle and associate professor Wayne Lichtenberger. But Project Genie was primarily led by a group of Berkeley students, some of whom were undergraduates. These included the future software pioneer L. Peter Deutsch, future Xerox PARC co-founder Butler Lampson, Chuck Thacker (a hardware specialist credited with the Ethernet, the laser printer, and the first computer that used a mouse), and Ken Thompson (co-creator of the Unix operating system). (Looking back on this illustrious group, all students of Dave Evans, Chuck Hutchins once told Professor Evans, "The only thing I missed in my life was being a student at Berkeley under your coaching.")

Project Genie took as its starting point a 24-bit computer from Scientific Data Systems (SDS) known as the SDS 930, one of the first computers with silicon-based transistors. Similar to the computers coming out of IBM at this time, the SDS 930 was housed in large metal cabinets. A well-equipped SDS 930 could easily exceed ten cabinets and require a 500-square-foot, climate-controlled room.

Using such a system, the Berkeley researchers modified the SDS 930 to provide protected memory and virtual memory, and then they wrote software for a timesharing operating system. In collaboration with SDS, the new version of this computer would be commercialized as the SDS 940.

It's hard to believe today that any *two* people could share this computer for anything useful, given its 192 KB of total system memory and 50 MB disk, with all the software written in assembly language. But Project Genie showed that the SDS 940 could be shared by up to six simultaneous users who could each have a maximum of 48 KBs of user memory at a time (swapping two 6 KB pages per revolution going to the fixed head-per-track drum that was spinning at 1,800 RPM). Each user worked within a protected space, so even if one user's program caused a crash, the other users wouldn't be affected.

Butler Lampson estimates that about sixty SDS 940s were eventually sold, mostly to companies that wanted to offer computer timesharing as a service, including two leaders in that business niche, Comshare of Ann Arbor and Tymshare of Palo Alto, California. The SDS 940 also stands out for its role in the creation of the internet when, in October 1969, an SDS 940 at Stanford University was connected, via the ARPA network, to a computer at UCLA.[3]

Computer timesharing was not, however, equivalent to the internet. It was more like the predecessor to today's cloud computing, combining networks of servers with remote access. Says Comshare co-founder Rick Crandall, "Timesharing was not just about the multi-user computer but also the network technology. The early networks of GE, Tymshare, and Comshare were not the peer-to-peer technology of the internet; rather, they were a 'star' system" (every host connected to the same central hub). "But still, the combo of the 940, the operating system, and the star network was definitely Cloud Computing 1.0."

Comshare concentrated computing power and then sold it to customers based on usage. With microprocessors and personal computers only a distant dream in the mid-1960s, entrepreneurs focused on a future world of work that depended on giant mainframe computers in central locations accessed by users with remote keyboards. That was Comshare's vision.

■ ■ ■

Rick Crandall started working in the U-M Computing Center (with Professor Frank Westervelt) in his sophomore year at U-M, while studying electrical engineering and mathematics. He was at the Computing Center in 1964 when a rep from Scientific Data Systems showed up, trying to sell the university on an SDS 930/940. "Nobody was interested...except me," Rick remembers. He tested out the timesharing claims of SDS by setting up a long-distance telephone line connection between the U-M Computing Center and Berkeley.[4]

"There was a cluster of professors, including Bernie Galler, who gathered around," Rick remembers. "The fact that I was typing in how to compute a simple least-squares algorithm interactively all the way to Berkeley and back was what impressed them."

Crandall was talking to Butler Lampson over the telephone throughout the test, because, as Crandall remembers, "Things were always going wrong. There were no hard disks at that time, so all of the file swapping was done on magnetic tape, which was grueling, and you kind of had to know when the system was down versus when it was waiting for a tape drive. So you needed the phone. It was a very manual process; however, the concept was very exciting, and the programmer productivity was clearly there. This was a way of getting multiple users online simultaneously using one machine. And so from my perspective the opportunity here was to try to make something commercial out of that."

Immediately after Comshare, Inc., was founded in Ann Arbor in 1966, Crandall moved to Palo Alto for a year to collaborate with UC–Berkeley, Tymshare, and SDS to enhance the potential of the SDS 940. They increased the system capacity to up to thirteen simultaneous users before they went their separate ways, with each organization taking a copy of the system they had all developed together.

Rick Crandall, co-founder of Comshare, seated at desk, with SDS 940 in background, circa 1967.
Photo courtesy of the Computer History Museum.

My wife doesn't understand me.
But my computer does.

ComShare the computer that understands

A Comshare advertising billboard posted on a Texas highway, circa 1968. Photo courtesy of the
Computer History Museum.

Crandall brought it back to Ann Arbor just as Bob Guise solidified the initial funding for Comshare. In the months and years to come, Comshare would develop a proprietary system to facilitate timesharing functions and manage the huge amount of data that were handled by the company's mainframes. (The fact that the SDS-940 had a 50 MB disk provides perspective on what was considered "huge" back then.) Meanwhile, Tymshare was doing the same thing.

■　■　■

As Comshare's CEO, Bob Guise was trying to figure out how to sell timesharing when he ran into Chuck Hutchins on that fateful spring day in 1967.

Chuck had become a small shareholder in Comshare while he was still at the Pontiac Varnish Company. He had spent evenings and weekends sitting at a card table in the corner of the family room with a teletype machine and an acoustic coupler (into which he placed the telephone handset), starting to learn what timesharing was all about. When he realized that the SDS 940 had more than three times the memory of the IBM 1620 (49,152 bytes instead of 15,000) and could hold almost ten times the number of instructions (16,384 instead of 1,666), he had an a-ha moment: "Perhaps we can use it to program NC machine tools." This was what he proposed to Bob Guise that day.

Guise remembers, "As we talked, we rapidly reached the conclusion that a numerical control application via timesharing, if it could be done under existing technology—which was not a given in May of 1967—would be a sure winner." It had now been eight years since Chuck had attended his first conference on numerical control, when the presenter guessed there were "hundreds" of numerically controlled machine tools in use. The industry had since responded enthusiastically to the potential for this technology, and by 1967, numerically controlled equipment accounted for perhaps a few thousand metal-cutting machine tools (nearly one-fifth of all such tools, according to a November 1967 University of Michigan presentation).[5] But given Chuck's experience at Buhr, he was certain that frustrations in programming these tools had increased along with demand.

Despite the potential of Chuck's idea, Comshare's budget for exploring new markets was tight. "Even with the new influx of capital," says Guise, "I had to be careful how we were spending money. Administration, R&D, marketing and sales, operations (facility expansion), and technical field support to include manuals were all competing vigorously for their share of the budget. The other obstacle was that within the computer science community, there were not all that many who knew what numerical control was in the first place. I sensed that I might encounter resistance from my own technical staff. They already had enough problems to solve." But Guise couldn't resist Chuck's enthusiastic certainty that numerically controlled machine tools were a huge untapped market for computer timesharing—if Chuck could figure out how to make it work.

As their impromptu lunch ended, Bob agreed that Comshare would provide sufficient office space and computer time to support the project, and would pay Chuck a living wage. "Bob shot from the hip and hired me," says Chuck, who took a substantial pay cut compared to what he'd been making at Pontiac Varnish, but the deal was better than Chuck could have hoped for: Suddenly, he had the support and resources he needed to pursue his idea. "Almost immediately I began writing the first 3,000+ lines of assembly language code for the SDS 940 that became the proof of concept."

6

Writing COMPACT

Chuck worked essentially alone for the next year. The team of highly competent programmers at Comshare didn't even know what an NC machine tool was, so they were little help. "To be honest," Bob Guise remembers, "over the next twelve to fourteen months, I was so embroiled with other problems that, other than an occasional 'Hello, how are you doing?' I didn't spend much time with Chuck's project." That was okay with Chuck, because he had a steep learning curve to climb.

"When I joined Comshare," says Chuck, "I was really a thirty-three-year-old pure techy. I knew nothing about business management or product marketing. And all I knew of computing I had learned on the job in a machine tool company. The only example of a programming language I'd seen was Hal Baeverstad's SPLIT. My plan was that I would write an equivalent to SPLIT for the SDS 940. After all, it had done the job for me at Buhr."

Chuck began by reading the manuals that Comshare was developing for its system. This introduced him to software and simple commands that had been developed by the collaboration with Tymshare and UC–Berkeley in California, like the Timesharing Operating System (OS), the Timesharing Assembly Program (TAP), the Digital Debugging Tool (DDT), and the Quick EDitor (QED). Right away, he realized that the SDS 940 was a binary machine, not a decimal digit machine like the IBM 1620.

Binary? What's that? Chuck thought. Without understanding binary numbers, he didn't even know how to write his code to represent ten inches, or how to compute a sine, cosine, or tangent. *And what are these floating point numbers?* He was told that the SDS 940 memory was divided into 8 pages of 2,048 24-bit words each, but he didn't know the best way to use each page. *And there's no subroutine library for functions in assembly language.*

In other words, everything was unfamiliar. With a little research and thinking through the implications for machine tools, Chuck came up with a plan. "The 1964 book *Digital Computing and Numerical Methods* bailed me out."[6]

First, he figured out what binary meant—the use of 1s and 0s that is now essentially synonymous with computing. "Once I understood binary," Chuck remembers, "I immediately taught my kids how to count in binary. That reinforced my own understanding. Then we got the idea to put birthday candles on the cake in binary. We would light the ones and not light the zeros. So, not long ago when I turned eighty, I only had to blow out two candles."

Next, he noted that machine tools have finite size, so he didn't need floating point numbers. He decided, "Let distances be a binary number, a sign bit and 23 bits is +/- 8,388,606. The majority of machine tools have inch input to four decimal places. Plus or minus 838 inches is a *big* machine, +/- 69 feet! Even a quarter of that is still a big machine, +/- 17 feet. So, I arbitrarily said, let's use two bits on the low end for rounding, the least significant bit would be .000025 inch. Then I looked at trig functions, sines and cosines. A sign bit, a zero or one, and a 22-bit binary fraction make the least significant bit equal to about a quarter of one millionth, almost equal to the Hans Hof seven place trig tables when I was manually programming. Good enough! I'm sure glad the SDS 940 had 24-bit memory; it seemed just enough to do this job with all fixed-point arithmetic."

Chuck knew he needed to be able to use the text editor (QED) to create both the software to program the SDS 940 and also to create the part programs to describe the machining of a part. So he studied how the QED created its memory layout—how it was started up, shut down, and so forth.

Following the convention of three-letter commands developed at Berkeley (like QED, TAP, DDT, and CAL), Chuck designated NCS (for Numerical Control System) as the command that would start a session associated with his project. "Again, take the easy stuff while you can," says Chuck.

And what to name his program? Chuck was always open about the fact that he basically copied and then adapted SPLIT, so he wrote to Sundstrand, asking for permission to call this version SPLIT and recognize Sundstrand and Harold Baeverstad for their contribution. Sundstrand turned him down with no explanation, but the company never questioned Chuck's use of their work. In fact, SPLIT had been copied once before, by Richard Stitt, founder of Numerical Control and Computer Services (NCCS) in Cleveland. Stitt called his adaptation— which worked on the IBM 1620 and later, the IBM 360—ACTION.

Because Chuck was now affiliated with Comshare, he decided to call his new program COMPACT, short for Comshare's Program for Automatically Controlling Tools. "Done; name chosen," says Chuck. "No bureaucracy when you are a one-man show."

As he grew in his understanding of how the eight pages of memory worked on the SDS 940, Chuck began to designate each page for a specific purpose. The pages were numbered ZERO through SEVEN. Page SIX was where the operating system performed a page-swap task; when the program called for a specific "syspage," the page would swap, and when the new page was in memory, execution began at the next sequential location. "We called this page the 'A' page," Chuck remembers, "and page SEVEN was the 'B' page. Later, page FIVE became the 'C' page." Page TWO contained canonical form data for point definitions, later extended to include lines, circles, and tabulated cylinders. Page THREE contained the COMPACT part program source. Page FOUR contained the formatted output manuscript. And NCS machine tool data were at the low memory end of Page ZERO.

Chuck's programming results in those first months included almost two pages of re-entrant, read-only code. "That was an important feature," he explains, "because when swapping pages, read-only pages didn't have to be written to the drum; they were already there to be read the next time it was your turn in memory. We were always thinking about efficiency."

With some of these most basic decisions made, Chuck created a COMPACT brochure that helped to articulate his vision, sales pitch, and services. The cover of the brochure offered this simple set of promises: "COMPACT is a user-oriented part programming language for the preparation of control tapes for numerically controlled machine tools. Designed to be used via remote teletype terminals, it is quickly learned and easily used. It saves programming time and prevents computational errors." Turning once again to the artistic skills of his wife, Ann, to illustrate the brochure, Chuck produced a document that he saw as his plan of action: "Implementing the code that would make what the brochure described a reality became my work objective."

By November 1967, Chuck had written enough code to participate in a seminar with U-M professor Frank Westervelt entitled "Demonstration of Numerical Control Programming Via Remote Computer Terminals." Their presentation was the final session of a two-day conference on U-M's North Campus that was all about numerically controlled machine tools. Other presenters included a representative of General Electric talking about ADAPT and someone from the Illinois Institute of Technology talking about APT. These were the unwieldy, mathematics-heavy program languages that had driven Chuck to SPLIT years earlier. While his early version of COMPACT could only program point-to-point instructions and didn't yet have a "machine tool link" to make it work on various kinds of machine tools, the men at the demonstration saw right away that COMPACT was easy to understand and use.

One of the first people to try it was a U-M industrial engineering student named Larry Schultz. He used COMPACT to program an NC lathe to produce a set of chess pieces. Schultz would go on to found Great Lakes Industry, Inc., a manufacturing company in Jackson, Michigan, that would become an early and longtime customer of MDSI.

■ ■ ■

After a year of working alone, Chuck was ready for some help. "There was no one in the Comshare R&D group who had any idea what I was doing," he says. "I had no one to discuss ideas with." With Bob Guise's blessing, Chuck called his former Buhr colleague Bruce Nourse. Bruce left Buhr and hired on at Comshare on April Fools' Day 1968.

The two men almost immediately left Ann Arbor for the annual conference of the Society of Automotive Engineers in Hackensack, New Jersey. Says Chuck, "We packed my Oldsmobile Vista-Cruiser with gear—teletypes, couplers, etcetera—and hit the road."

In fact, Chuck picked up Bruce from Buhr at the end of his last day of work there. "I had my luggage at work," says Bruce, "and I never went back to my apartment until after our trip." They drove straight through the night, arrived at Comshare's office in Hackensack before dawn, set up the computer, and got to work. They spent the next thirty-six hours refining the COMPACT code to ensure a successful demo.

On Sunday evening, they were checking into their hotel near the conference when another Ann Arbor tech entrepreneur, Chuck Newman (who would eventually found ReCellular, Inc.), appeared in the lobby without a room. Chuck and Bruce invited him to get a cot and share theirs. "On Monday morning, bright and early," Chuck remembers, "we bought a two-wheeled pushcart for $14.95 to haul all our gear out of the hotel without having to tip a bellhop. Anything to save Bob Guise's Comshare dollars."

They drove home as soon as the conference was over. Bruce chuckles at the memory: "We always said we drove to Hackensack 'n' back."

After that trip, Bruce began to scrutinize Chuck's yearlong programming effort. As Bruce remembers it, "Chuck headed out on vacation, but before he left, he gave me a stack of papers three-quarters of an inch thick and said, 'Here's what I have, and I want you to add this, this, this, this, this, this, this, and this.'" Working out of a second bedroom in the Ann Arbor apartment he shared with his new wife, Bruce tackled the changes. "When Chuck got back, I had about two inches worth of source code listing, and I had put in all the features that he wanted to go out and demonstrate."

Bruce recommended a lot of changes, most of which made sense to Chuck, who remembers, "I successfully defended some decisions about how to do things, but his superior skill triumphed in many cases. The amazing thing is that all my schemes for storing data and all the function routines survived." With that, Bruce took over the software effort for COMPACT, and, says Chuck, "I haven't actively written any software since." Bruce became the "lead system guru" and set the example for those who would join the R&D group down the road.

That experience also set the tone for Chuck and Bruce's partnership for decades to come. "Chuck is a visionary," says Bruce. "He has all

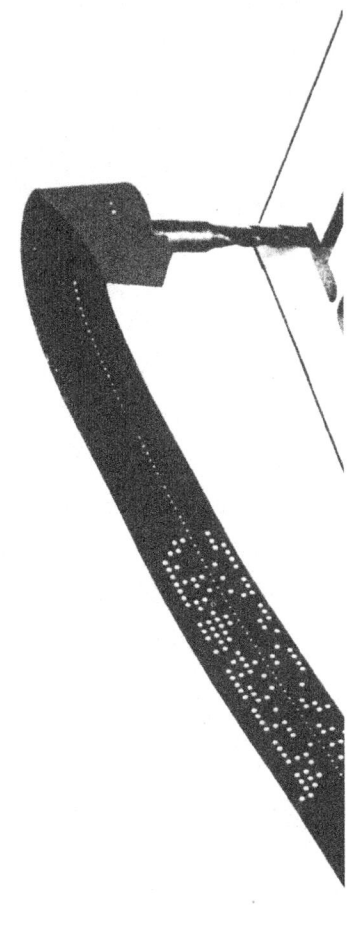

Com-Share
INCORPORATED

COM-SHARE'S **P**ROGRAM FOR **A**UTOMATICALLY **C**ONTROLLED **T**OOLS.

The first COMPACT brochure, written by Chuck Hutchins, with illustrations by Ann Hutchins.

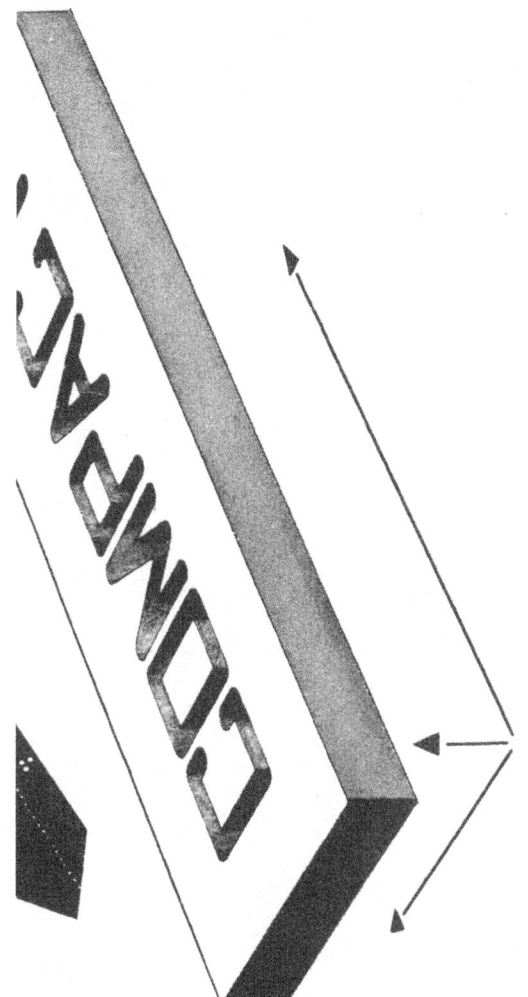

COMPACT is a user-oriented part programming language for the preparation of control tapes for numerically controlled machine tools. Designed to be used via remote teletype terminals, it is quickly learned and easily used. It saves programming time and prevents computational errors.

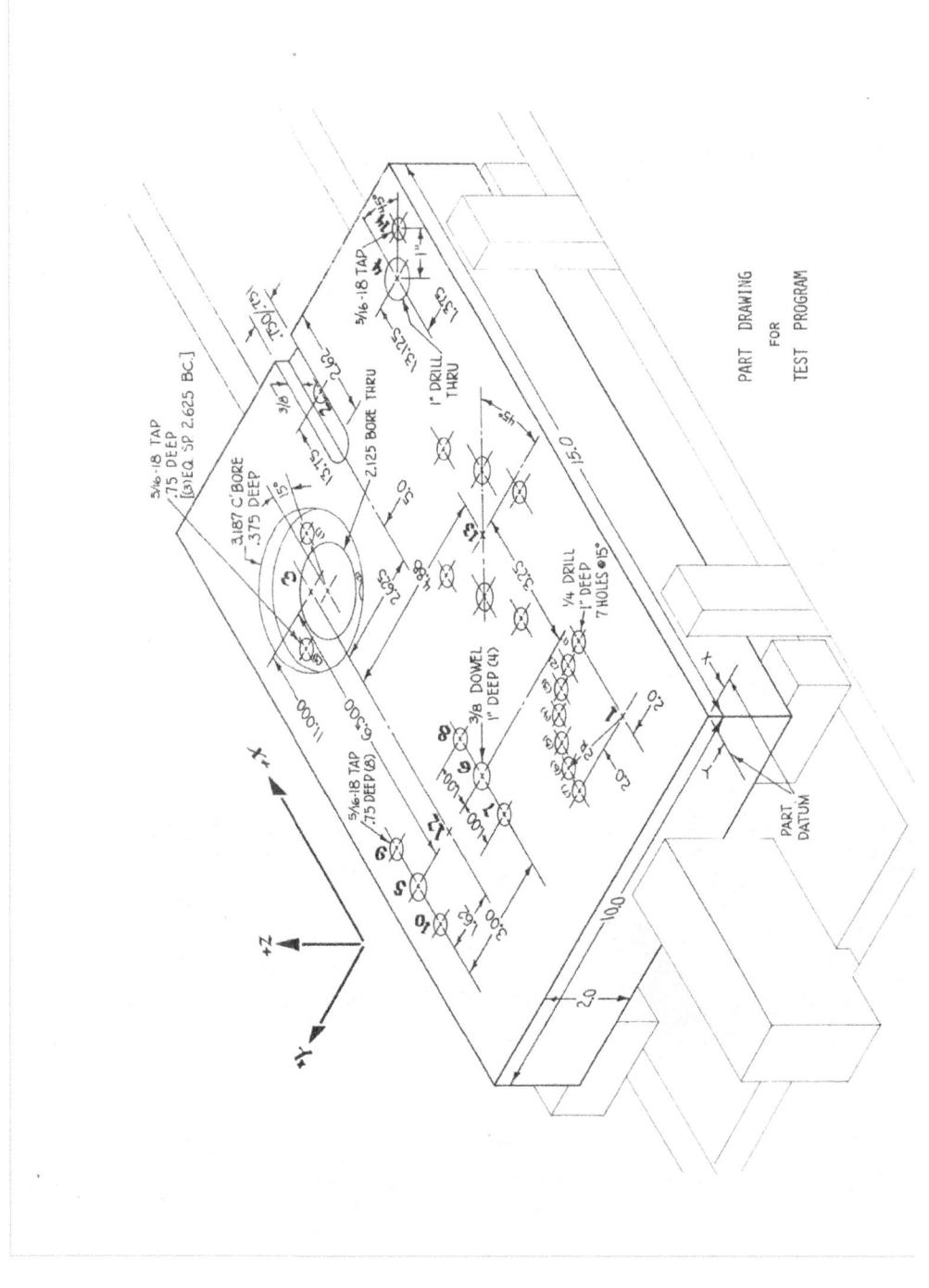

PART DRAWING
FOR
TEST PROGRAM

HOT TECH COLD STEEL

```
IDENT, TEST PROGRAM
SETUP, 60X, 0Y, 12Z

*BASE FOR NESTING BLOCKS AND PARALLELS
BASE, 3XB, 10YB, 1.5ZB

*BASE FOR PART RELATIVE TO NEST
BASE, 0XB, 0YB, 2.1ZB

*SET PT 100 EQUAL TO PRESENT BASE
DPT 100, 0XB, 0YB, 0ZB
DPT 1, 2XB, 2YB, 0ZB
DPT 2, 13.75XB, 5YB, 0ZB
DPT 3, 11XB, PT2, 2.625Y
DPT 4, 13.125XB, 1.375YB, 0ZB
BASE, PT4
DPT14, PT4, 1X, ROTXY45
BASE, PT 100
DPT 12, PT 3, -6.5X

*(4) SCREWS AND (2) DOWELS IN PATTERN
STGRP1, 5, 6
DPT 5, PT 12, 1.62Y
DPT 6, PT5, -3Y
STGRP2, 9, 10, 7, 8
DPT 9, PT5, 1X
DPT 10, PT 5, -1X
DPT 7, PT10, -3Y
DPT 8, PT9, -3Y
DPT 13, PT12, 3.25X, -4.88Y

*USE 2" DRILL AT PT 3
ATCHG, TOOL1, GLI5.5, TD2, 80FPM, .024IPR
DRL, PT3, 2.8A

*USE 1 1/4 6 FLUTE END MILL TO ROUGH C BORE
ATCHG, TOOL2, GLI2.4, TDI.25, 70FPM, .021IPR
MOVE, PT 5, .3X
MOVE, -.43ZB
ICON360, PT3, -.43ZB, 3.1BC, OCW, TOL.01
MOVE, PT3

*USE 5/8 TWO FLUTE END MILL TO ROUGH KEY
ATCHG, TOOL3, GLI1.7, TD.625, 70FPM, .012IPR
MOVE, PT 2, 15.4XB
MOVE, -.43ZB
CUT, -3LX
ICON180,PT2,12.38XB,-.43ZB,90CW,.740BC,TOL.01
ICON180,PT2,15.4XB,-.43ZB,270CW,.740BC,TOL.1
CUT, 12.38XB
MOVE, .01Y, 0ZB

*USE "F" DRILL FOR 5/16-18 TAPPED HOLES
ATCHG, TOOL4, GLI4, TD.25, 80FPM, .005 IPR
DRL3, PT3, 2.625BC, .37RA, 1.2A, 15CW
DRL, 1.2A, PTGRP2
BASE, PT13
DRL, 1.2A, 3.25X, -4.88Y, ROTXY-45, PTGRP2
DRL, PT14, 1.2A

*USE 1/4 DRILL FOR 7 HOLES IN QUADRANT AT PT 1
ATCHG, TOOL5, GLI4, ID.25, 80FPM, .005IPR
DRL24, PT1, 2R, OCCW, 1.25A, SKP-8

*USE 11/32 DRILL FOR (4) 3/8 DOWELS
ATCHG, TOOL6, GLI5.1, TD.34, 80FPM, .007 IPR
DRL, 1.2A, PTGRP1

*BASE IS STILL AT PT 13 FOR THE ROTATION WHICH FOLLOWS.
DRL, 1.2A, 3.25X, -4.88Y, ROTXY-45, PTGRP1

*USE 1" DRILL AT PT 4
ATCHG, TOOL9, GLI6, 1TD, 70FPM, .016IPR
DRL, PT4, 2.5A

*USE 1 1/4 6 FLUTE END MILL TO FINISH C'BORE
*LEAVE FINISH BORING STOCK ON OUTSIDE DIAMETER.
ATCHG, TOOL7, GLI2.4, TDI.25, 90FPM, .015IPR
MOVE, PT 3, .3X
MOVE, -.475ZB
ICON360, PT3, -.475ZB, 3.18BC, OCW, TOL.002
MOVE, PT3

*USE 5/8 (4) FLUTE END MILL TO FINISH KEY
ATCHG, TOOL8, GLI1.7, TD.625, 90FPM, .012IPR
MOVE, PT 2, 15.4XB
MOVE, -.475ZB
CUT, -3LX
ICON180,PT2,12.38XB,-.475ZB,90CW,.7505BC,TOL.00
ICON180,PT2,15.4XB,-.475ZB,270CW,.7505BC, TOL.1
CUT, 12.38XB
MOVE, .01Y, 0ZB

*USE 5/16-18 TAP FOR HOLES IN GROUP 3
ATCHG, TOOL10, GLI8, TD.31, 65FPM, 18PITCH
FLT3, PT3, 2.625BC, .37RA, .7A, 15CW
FLI 14, PTGRP2
BASE, PT13
FLI, .7A, 3.25X, -4.88Y, ROTXY-45, PTGRP2
FLI, PT14, .7A

*USE 3/8 TWO FLUTE END MILL FOR 3/8 DOWELS
ATCHG, TOOL11, GLI2, TD.37, 80FPM, .007 IPR
DRL, 1.1A, PTGRP1

*BASE IS STILL AT PT 13 FOR THE ROTATION WHICH FOLLOWS.
DRL, 1.1A, 3.25X, -4.88Y, ROTXY-45, PTGRP1

*USE 2.125" MICRO BORE BAR FOR HOLE AT PT 3
ATCHG, TOOL12, GLI4.5, TD2.12, 200FPM, .005IPR
BORE, PT3, .36RA, 1.82A

*USE 3.187" MICRO BORE BAR FOR C'BORE AT PT 3
ATCHG, TOOL 13, GLI5.5, TD3.187, 200FPM,.005IPR
BORE, PT3, .470A
END
```

MAJOR OPERATIONS

Major operations transfer information to the computer or instruct it and ultimately the machine tool to take some action. They must occur one per statement. Their relative position is not important, but is generally first for clarity.

IDENT   IDENTify causes the characters following the comma to be punched into the leader of the tape in a manner easily read by the people involved.

SETUP   SETUP defines the coordinate location of the machine tool slides at the start of the part program. It may include a limitation on the rapid traverse IPM if this is possible on the machine.

BASE   BASE is the reference point for all XB, YB and ZB dimensions. This point is initialized to absolute zero and may be moved at the convenience of the part programmer.

DPT n   Define PoinT permits the definition of a point using many reasonable combinations of minor operations and stores the absolute coordinates for later retrieval at PT.

DMP n   Define Machine Path permits the definition of cutting paths by the lengths of their respective relative coordinate vectors LX, LY and LZ and stores these for later retrieval as MP or -MP.

ATCHG   Automatic Tool CHanGe causes the machine tool to change tools automatically if it is capable of doing so. For the convenience

CUT   CUT is the same as MOVE but at feed rate. In addition, a CUT statement which includes an absolute coordinate and a path will MOVE (at rapid rate) to the coordinate point and then CUT the path at feed rate.

The following major operations are specialized combinations of MOVE and CUT with all the features described above. Each may be used by itself to perform a single operation or with n to perform n equally spaced operations on a line or bolt circle. The operations may be performed at many defined locations through the use of the minor word PTGRP. Each requires "A" for Advance (the cutting stroke) and may have "RA" for Rapid Advance prior to cutting. Note: If *n* is omitted, one is implied for these three major operations only.

DRL *n*   MOVE to location, RA if given, feed in a distance A and rapid return to the initial location.

BORE *n*   BORE is the same as DRL except feed out the distance A instead of rapid traverse.

FLT *n*   FLoat Tap is the same as bore except feed rate is determined by RPM-PITCH combination and the spindle is reversed at the end of the feed stroke.

The last three major operations pertain to moving the tool point in a path approximating a circle. This applies primarily to machines with contouring controls, but may prove useful on point-to-point machines for some operations.

CONT *n*   CONTour *n* degrees of a circle. The tool point will CUT to the start of the circle and CUT around it as described by the minor operations. This is a chordal approximation with the number of chords determined by TOL.

HOT TECH COLD STEEL

ICON *.n.*
OCON *.n.*
Inside CONtour and Outside CONtour are variations of contour permitting tool diameter compensation either inside or outside of the diameter specified. This "tool offset" is especially useful when positioning mills for cutting with the side. Changes in tool diameter (TD) in a tool change statement will result in correct tool positioning wherever ICON or OCON are used.

MINOR OPERATIONS

Minor operations are used to supply the parameters necessary to make the major operations function as desired. There is a great deal of flexibility in their use and each completed statement results from a logical combination of a major operation and several minor operations.

Dimensioning Related Minor Operations

XB n
YB n
ZB n
XBase (etc.) dimensions are relative to the current base and take precedence over all other means of obtaining a coordinate location.

X n
Y n
Z n
X (etc.) dimensions are delta type dimensions and are used to modify a previously defined point or the current tool point location. NOTE: It is not permissable to use X with XB in the same statement.

LX n
LY n
LZ n
ALong X (etc.) dimensions define paths for cutting, or lines for drilling equally spaced holes. They can be used with either X or XB (etc.) dimensions. When used with CONT they cause a linear displacement in each chord. Thus, LZ used with CONT would cause the tool point to follow a helical path where the LZ dimension is the pitch.

of restarting the machine tool in the middle of a program, this occurs at the nearest even X-Y inch in the direction of the setup point.

MTCHG
Manual Tool CHanGe causes the machine tool slides to return to the setup point and stop for manual tool change.

TCHG
Tool CHanGe causes the machine tool slides to go to the X-Y coordinates specified in the statement, and stop with a fully retracted Z axis. This is useful for clearing the part for inspection or manual tool change.

STORE n
STORE causes the remainder of the statement following the comma to be stored for later retrieval.

STGRP n
STore GRouP stores point numbers or path numbers as a group in the order in which they are to be executed. They can be retrieved as point numbers with PTGRP n or as path numbers with MPGRP n. Negative path numbers may not be used.

END
END terminates part program, returns the machine tool axes to the setup point, returns tool one to the spindle if the machine has an automatic tool changer, and rewinds tape.

All remaining major operations cause the tool point to perform the activity specified. Each motion will be checked for exceeding the machine tool limits, and on vertical machines, the tool point will be checked for damaging the table. Other checking may be implemented in the future.

MOVE
MOVE causes the tool point to go to a coordinate location at rapid traverse rate.

MP n — Machine Path retrieves from memory the path distances created by the DMP major operation.

XSTOR n — XSTOR is used to retrieve from storage a partial or complete compact statement (XSTOR2) or several sequentially stored statements (XSTOR2.05) the latter being statements two through five.

PTGRP n — PoinT GRouP — To retrieve a sequence of numbers previously stored as a STGRP and execute these as point numbers for performing operations at many predefined locations.

MPGRP n — Machine Path GRouP — Same as PTGRP, but cuts a sequence of predefined paths.

Minor Operations for Machine Control

ACCEL DECEL — ACCELeration and DECELeration force the insertion of these codes into tape blocks defining cutting motions where they would not normally appear. They are used for the smooth starting and stopping (no overshoot) of high speed cuts, and are not necessary at normal feed rates.

STOP — STOP generates EIA M00 code to stop all functions of the machine tool at the end of the current operation.

OSTOP — Optional STOP generates EIA M01 code to stop all functions of the machine tool if the machine operator has preset the optional stop logic of the controller.

CON n — Coolant ON — turns on coolant. If more than one type is available, a number will select a specific one.

LR n — ALong R is a special path to permit contouring a circle with a uniformly increasing radius (-LR causes uniform decrease) or for drilling holes with equal angular spacing on a bolt circle of increasing (or decreasing) radius.

RA n — Rapid Advance is a delta Z motion of the tool toward the work at rapid traverse rate.

A n — Advance is a delta Z motion of the tool into the work at cutting feed rate.

R n — Radius of the circle to be contoured or drilled.

BC n — Bolt Circle is the same as twice R.

TOL n — TOLerance is the distance from the chord to the true circle when contouring. It is used to determine the number of chords.

CW n CCW n — ClockWise and CounterClockWise determine the direction of drilling or contouring. The number indicates the starting angle (degrees) which must be measured from zero at the plus X axis and in the direction to be contoured or drilled. Holes may be thought of as being numbered sequentially in this direction and from this initial point.

ROTXY n — ROTate in the XY plane. Rotation occurs about the present base if a point, or base dimension is used. Rotation is about the present tool point for X or LX (etc.) dimensions. CCW is minus.

Minor Operations for Spindle Speed Control

RPMRH n — Revolutions Per Minute Right Hand starts the spindle in the right-hand or clockwise direction.

RPMLH n — Revolutions Per Minute Left Hand starts the spindle in the left hand or counterclockwise direction.

TD n — Tool Diameter - the cutting diameter of the tool. Used for tool offset and spindle speed computation.

FPM n — Feet Per Minute - peripheral velocity of the cutting tool. Must be used with TD.

NOTE: The spindle will be started with TD and FPM. The rotation will be right hand unless RPMLH is included.

Minor Operations for Feed Rate Control

IPM n — Inches Per Minute cutting feed rate.

IPR n — Inches Per Revolution permits computation of inches per minute feed rate using the current RPM.

PITCH n — PITCH permits computation of feed rate for tapping.

Minor Operations for Tool Changing

TOOL n — TOOL number for selection of tools if a changer is available.

GL n — Gage Length - The length of the tool from the gaging point in the spindle to the cutting point.

Minor Operations for Data Retrieval

PT n — PoinT retrieves from memory the absolute coordinates created by the DPT major operation.

COF — Coolant OFf.

DWELL *n* — DWELL, for machines with this capability, causes a pause of 0.3 seconds to occur. If a number is included, this will be the DWELL time in seconds.

RDWEL n — Revolution DWELl – As above, causes a timed dwell, but the duration of the dwell will be computed to permit the spindle to make the prescribed number of revolutions before continuing. NOTE: Not all controllers are equipped to handle DWELL and RDWEL.

Miscellaneous Minor Operations

SKP n — SKiP Hole. When drilling equally spaced holes on a line or circle, the holes may be considered to be numbered sequentially from one to the total number requested. Any of these holes except the first may be omitted with this command. It is valid for numbers from 2 to 263. A negative skip hole number will eliminate that hole and any remaining holes in the pattern. Thus, for example, 11 equally spaced holes in a quadrant could be drilled by asking for 40 holes on the circle and using SKP-12.

BD — Block Delete. Puts a code in each block of tape generated to permit parts of the program to be py-passed at the discretion of the machine tool operator. See complete manual for full description.

INSRT — INSeRT puts actual characters directly on the tape. The computer will be unaware of any activity caused by this command. Be cautious with its use.

MANY ADVANTAGES..

COMPACT, offered as a Numerical Control Subsystem (NCS) on Com-Share's time-sharing computer, will allow manufacturers to employ computer-assisted parts programming, an advantage previously reserved for companies large enough to own an *"in house"* computer.

COMPACT will be extended with plane geometry capabilities, permitting the definition and storage of line and circle data and the computation of related intersections and tangents. These capabilities will be similar in many respects to the geometry capabilities of the ADAPT machine tool program, but will be in a free format, as are all COMPACT statements.

COMPACT understands a vocabulary of sixty-two symbols . Nineteen of these, used one per statement, describe the

HOT TECH COLD STEEL

kind of operation being performed. Forty-three symbols describe minor operations and are used to supply the parameters necessary to execute the major operation.

COMPACT commands consist of symbols (*five letters or less*) or combinations of symbols and numbers with sign and decimal point if necessary. Numbers with a decimal fraction will be extended to four places by the program. Each symbol, or symbol and number, is separated from the next by a comma. Each complete statement consisting of a logical combination of symbols and numbers is terminated by a carriage return.

COMPACT will accept the following characters:

A through Z
0 through 9
Plus (+) indicates positive numbers but is not required.
Minus (-) indicates negative numbers.
Comma (,) separates the parts of a COMPACT statement.
Blank improves readability but is ignored by COMPACT.
Line Feed permits the continuation of a COMPACT statement to the next line.
Carriage Return terminates all COMPACT statements.

In the language description which follows, symbols which require a number to define the operation will be followed by "n". If the number is optional, *n* will be used. When the number is omitted, zero is implied except for the commands DRL, FLT, BORE, where one is implied.

COMPACT, enables a machinist or a process engineer who understands conventional machine tools and cutting operations to sit at the typewriter keyboard of a Com-Share teletype in his own office and type out directions to control a machine tool after only a few hours of training. The symbols and directions he must type are quickly learned. Com-Share's time-sharing computer then turns his directions into machine control tapes in minutes.

COMPACT automatically checks his directions and points out grammatical errors immediately, permitting on-the-spot editing. In addition, it will warn him when he attempts to program the tool to go beyond its dimensional limits.

COMPACT will continue to grow in a way that best suits the needs of its users. Ideas will be shared and the future of the program and language development will be guided through an N/C Users Group. Additions and deletions can be easily made as the users' requirements are determined.

COMPACT commands are available to permit operations to be executed at patterns of defined coordinates or equally spaced on lines, circles, or spirals. A single COMPACT statement will initiate these commands to control the simplest drilling operation or complex sequences of cuts with the same tool. The linear translation and/or rotation of these pattern operations is also possible.

THE MOST SOPHISTICATED SOFTWARE, ANYWHERE...

The COM-SHARE time sharing system is a unique combination of hardware and COM-SHARE's own software especially configured for conversational usage from a remote terminal.

The following are the most important of the subsystems making up the COM-SHARE system. For more detailed information on software, write for our brochure "COM-SHARE CAPABILITIES".

CONVERSATIONAL FORTRAN IV. A conversational language for general computer projects. Instant response to each statement. Statements can be changed without recompiling the entire program.

FORTRAN II—a highly efficient time sharing version of full FORTRAN II including a run time environment which interacts with the user on an instantaneous basis.

CAL. For numerical problems is highly interactive environment. Relieves user of all burdens of storage allocation for both programs and data.

BASIC. Similar to FORTRAN, but much easier to use.

QED. A generalized text editor to create and modify symbolic programs, data or text for any purpose.

HELP. A direct self-teaching facility. Accepts questions on system or subsystem usage in natural language and answers in English.

DDT. A debugging package providing examination, search, change, and insertion of break-point and step-trace instructions.

SNOBOL. Provides complete facilities for manipulation of character strings.

TAP. A two-pass machine language assembler with sub-program, literal, and macro facilities.

LIBRARY. The COM-SHARE system includes a wide selection of library programs available to all users of the system. Continual updates and additions to the library make it an invaluable feature of the system.

AN EXPANDING NETWORK OF TIME SHARING FACILITIES...

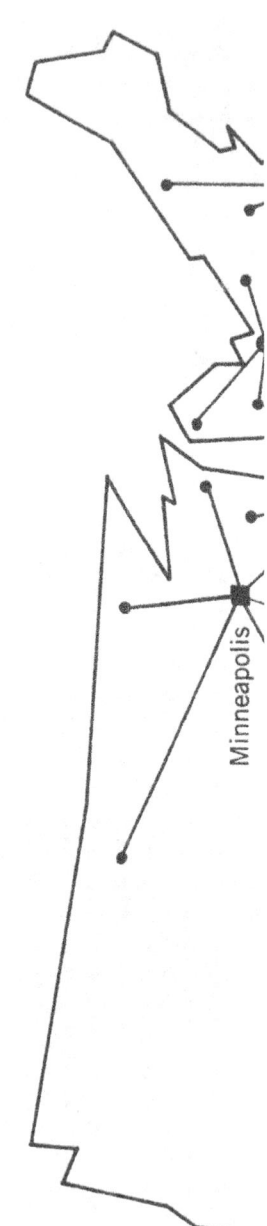

Minneapolis

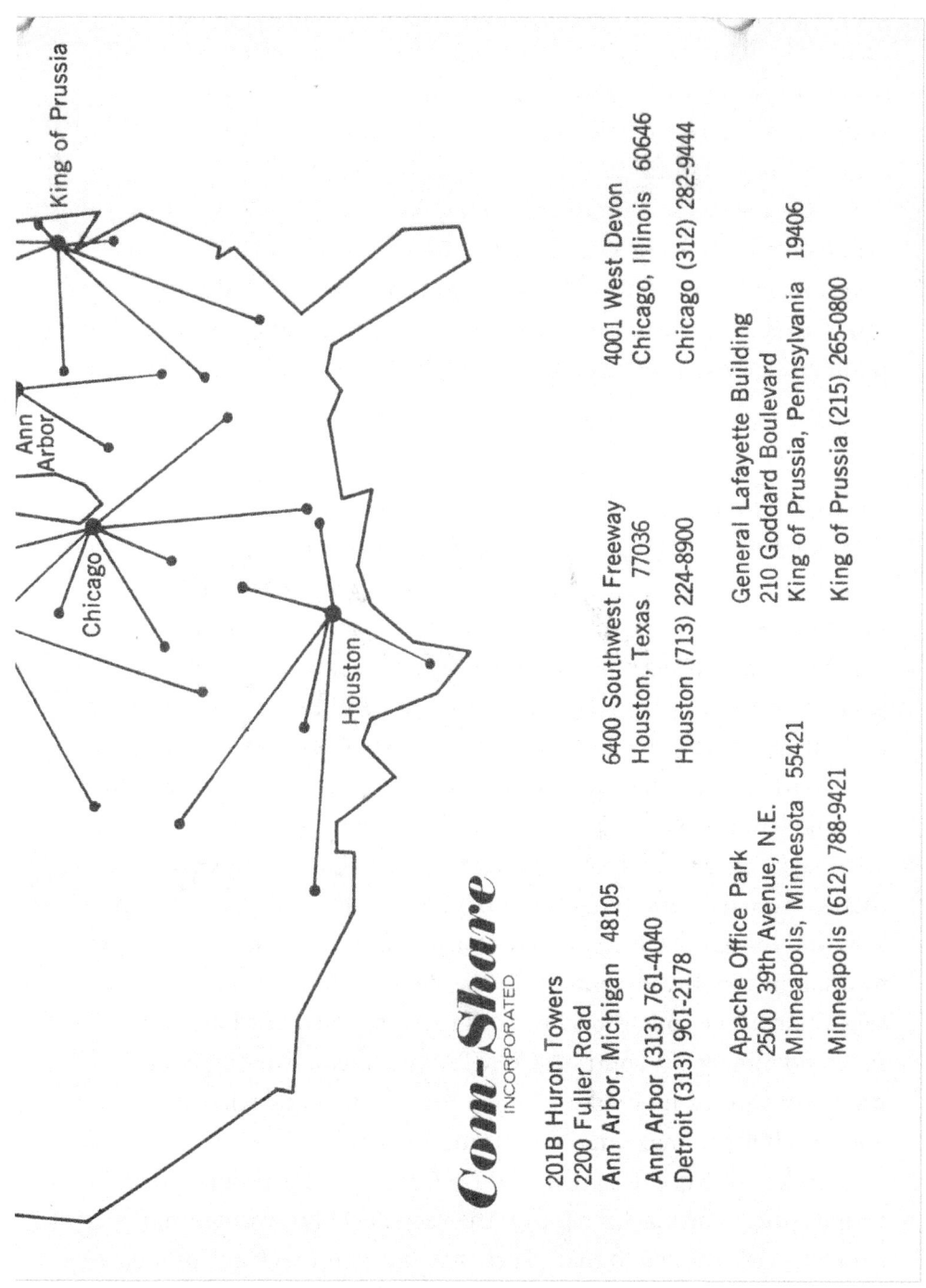

Com-Share
INCORPORATED

201B Huron Towers
2200 Fuller Road
Ann Arbor, Michigan   48105

Ann Arbor (313) 761-4040

Detroit (313) 961-2178

6400 Southwest Freeway
Houston, Texas   77036

Houston (713) 224-8900

4001 West Devon
Chicago, Illinois   60646

Chicago (312) 282-9444

Apache Office Park
2500 39th Avenue, N.E.
Minneapolis, Minnesota   55421

Minneapolis (612) 788-9421

General Lafayette Building
210 Goddard Boulevard
King of Prussia, Pennsylvania   19406

King of Prussia (215) 265-0800

kinds of ideas, and he sees all kinds of possibilities. I'm the person who hears his idea, and I say either, 'Yeah, that's a great idea,' and then I do it, or I tell him why it won't work. If I tell him it won't work, but it's something he really wants to do, he'll rattle around trying to come up with a different way to get there. If he runs out of ideas but still really wants it, then it's time for me to start getting creative."

Bruce continues, "Programming is a very inventive thing. There's an art involved. There are a million ways to do something in a computer program, but only a few of them are good. Whatever you do, you want a program that is bug-free, won't crash on customers, and will be easy to maintain for years and years."

■ ■ ■

One of the functionalities that Chuck and Bruce improved together was COMPACT's "single-pass" processing, through which each statement was processed completely before the next was begun. Before COMPACT, most NC programs used batch processing. Chuck compares batch processing to launching a rocket with no ability to control its course: "The parts program was submitted to the computer and the results returned were either diagnostic messages—errors that needed to be fixed—or successful results, without any assurance which it would be." Programs with errors were revised and processed again, and maybe again after that. Add to this the need for a "post processor" that converted a basic program into instructions for a specific machine tool, and the approach could take from twenty-four to forty-eight hours to result in a tool-ready punched paper tape. "From a cost standpoint," says Chuck, "the sporadic nature of success resulted in a lot of wasted time and money." In contrast, COMPACT's interactive, single-pass processor worked more like a "moon rocket with good telemetry data and the ability to make mid-course maneuvers."[7]

Chuck and Bruce designed COMPACT so that the programmer would specify the machine tool at the very start of the program. Bruce explains, "We wanted to make sure that any commands the user gave were within the travel limits of the machine and the feed limits of the machine and the rapid limits of the machine as the program was being written."

They still had room on Pages ZERO and ONE of the eight pages of memory to specify the machine tool. Chuck remembers, "All of our program data were in the lower half of page ZERO. We decided we could put 'post processor' instructions in the upper half of page ZERO, and we could bring in page ONE with supporting code for classes of machine tools. So all the drilling machines had a separate page ONE, all the turning machines had a different page ONE, and so on. That gave us the flexibility to do the post processing job *not post*, but simultaneously." They called these specialized instructions "machine tool links" or just "links."

With the machine defined upfront, the computer was able to recognize right away if any instructions typed by the part programmer were inconsistent with the intended machine tool. The computer would flag the error immediately, enabling the part programmer to correct his program as he created it. Later promotional materials summed it up this way: "In this manner, the program proceeds to completion resulting in a completely debugged program during the 'single pass' through the computer."[8] Chuck would describe the programming experience with COMPACT as carrying on a conversation with a computer, rather than writing letters back and forth.

Bruce continued revising COMPACT throughout the summer of 1968. He edited code late into the night and started the assembly process and system test running on his home teletype at seven in the morning. "That's when I would shave and shower and eat my breakfast," says Bruce, "and then I'd continue working on the software until the machines shut down at eleven at night. It was a lot of long hours."

When Bruce and Chuck had completed their refinements, the newly revised program earned a revised moniker: COMPACT II. The distinguishing feature of COMPACT II from the original version was the ability to define lines and circles and do some planar geometry. With that, they had a program that would be vastly more useful to customers.

COMPACT II would be described as "a unique, user-oriented, conversational, interactive, time-shared, single-pass, numerical control processor, the output of which is capable of operating 2-, 3-, and 4-axis machine tools. Extensive 2-axis geometry permits virtually all line, circle, intersection, and tangent computations."[9] With the software now optimized for commercial success, it was time for Chuck to don a new hat: lead salesman.

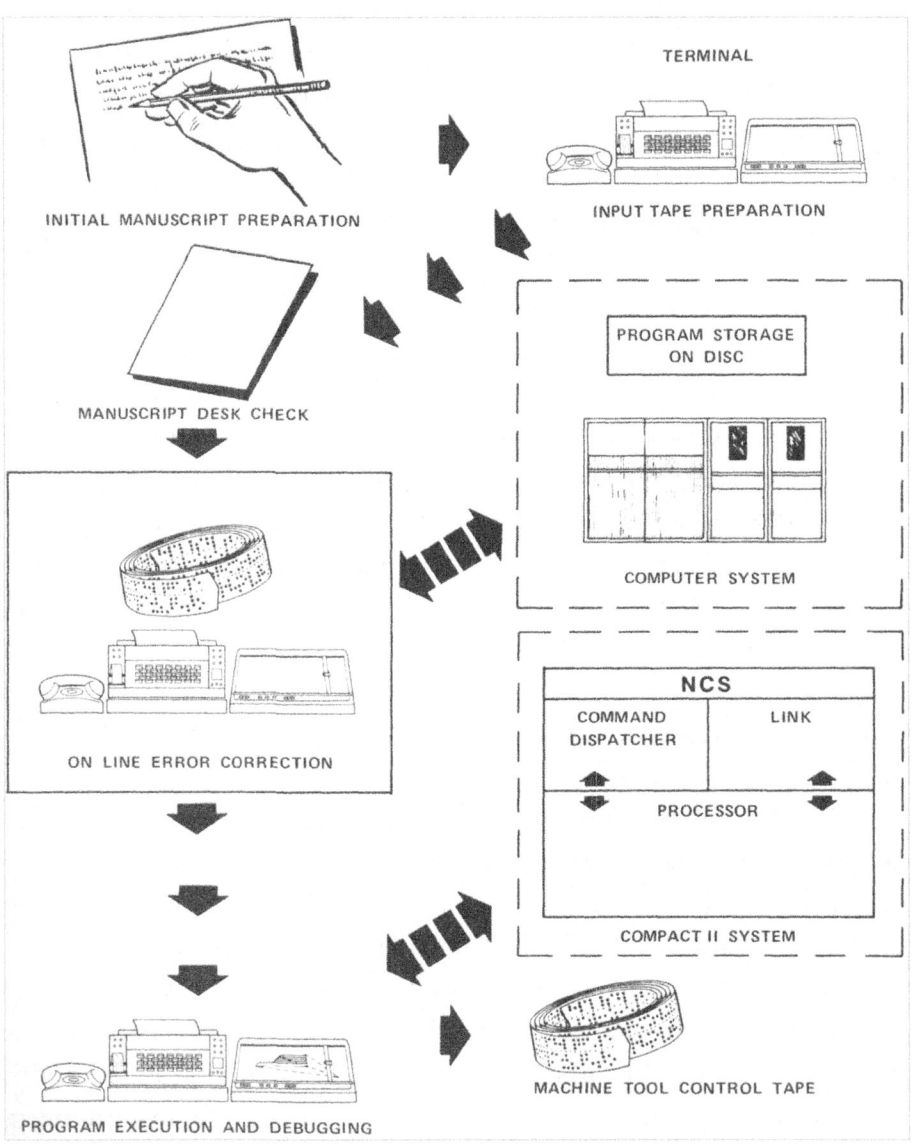

Typical process used by a part programmer to create a machine tool-ready punched paper tape using COMPACT II.

MANUFACTURING DATA SYSTEMS, INC.

presents

COMPACT II

COMPACT II is a unique, user-oriented, conversational, interactive, time-shared, single-pass N/C processor, capable of operating two, three, and four axis machine tools.

MDSI COMPACT II brochure.

Drilling

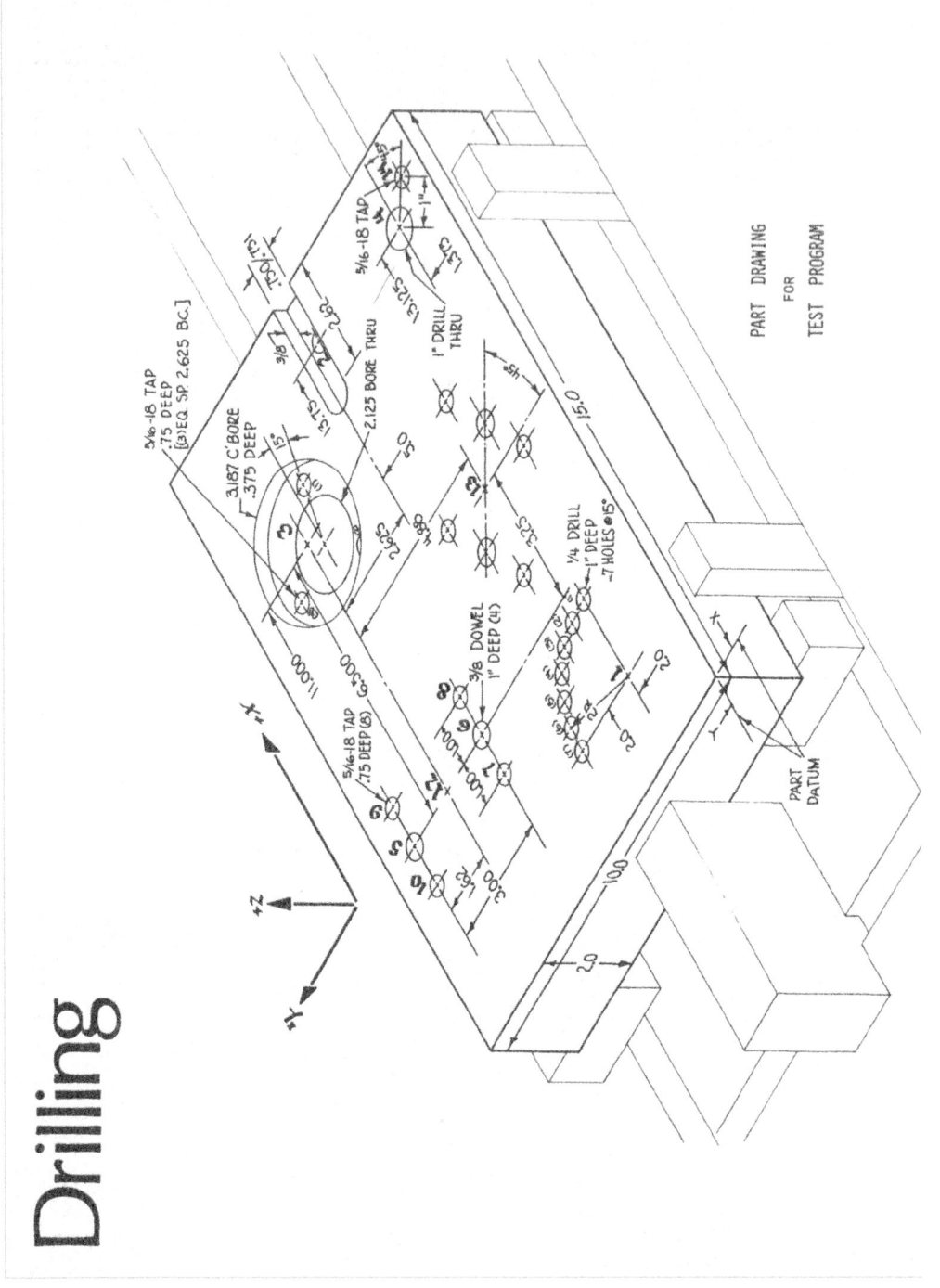

PART DRAWING
FOR
TEST PROGRAM

HOT TECH COLD STEEL

```
IDENT, TEST PROGRAM
SETUP, 60X, 0Y, 12Z

*BASE FOR NESTING BLOCKS AND PARALLELS
BASE, 3XB, 10YB, 1.5ZB

*BASE FOR PART RELATIVE TO NEST
BASE, 0XB, 0YB, 2.1ZB

*SET PT 100 EQUAL TO PRESENT BASE
DPT 100, 0XB, 0YB, 0ZB
DPT 1, 2XB, 2YB, 0ZB
DPT 2, 13.75XB, 5YB, 0ZB
DPT 3, 11XB, PT2, 2.625Y
DPT 4, 13.12XB, 1.375YB, 0ZB
BASE, PT4
DPT14, PT4, IX, ROTXY45
BASE, PT 100
DPT 12, PT 3, -6.5X

*(4) SCREWS AND (2) DOWELS IN PATTERN
STGRP1, 5, 6
DPT 5, PT 12, 1.62Y
DPT 6, PT5, -3Y
STGRP2, 9, 10, 7, 8
DPT 9, PT5, IX
DPT 10, PT 5, -IX
DPT 7, PT 10, -3Y
DPT 8, PT9, -3Y
DPT 13, PT12, 3.25X, -4.88Y

*USE 2" DRILL AT PT 3
ATCHG, TOOL1, GL15.5, TD2, 80FPM, .021PR
DRL, PT3, 2.8A

*USE 1 1/4 6 FLUTE END MILL TO ROUGH C BORE
ATCHG, TOOL2, GL12.4, TD1.25, 70FPM, .021PR
MOVE, PT 3, .3X
MOVE, -.43ZB
ICON360, PT3, -.43ZB, 3.18C, OCW, TOL.0!
MOVE, .01Y, 0ZB

*USE 5/8 TWO FLUTE END MILL TO ROUGH KEY
ATCHG, TOOL3, GL11.7, TD.625, 70FPM, .0121PR
MOVE, PT 2, 15.4XB
MOVE, -.43ZB
CUT, -3LX
ICON180, PT2, 12.38XB, -.43ZB, 90CW, .740BC, TOL.01
ICON180, PT2, 15.4XB, -.43ZB, 270CW, .740BC, TOL.!
CUT, 12.38XB
MOVE, .01Y, 0ZB

*USE "F" DRILL FOR 5/16-18 TAPPED HOLES
ATCHG, TOOL4, GL14, TD.25, 80FPM, .005 IPR
DRL3, PT3, 2.625BC, .37RA, 1.2A, 15CW
BASE, PT13, PTGRP2
DRL, 1.2A, PTGRP2
DRL, 1.2A, 3.25X, -4.88Y, ROTXY-45, PTGRP2
BASE, PT13
DRL, PT14, 1.2A

*USE 1/4 DRILL FOR 7 HOLES IN QUADRANT AT PT 1
ATCHG, TOOL5, GL14, TD.25, 80FPM, .0051PR
DRL24, PT1, 2R, 0CCW, 1.25A, SKP-8

*USE 11/32 DRILL FOR (4) 3/8 DOWELS
ATCHG, TOOL6, GL15.1, TD.34, 80FPM, .007 IPR
DRL, 1.2A, PTGRP1

*BASE IS STILL AT PT 13 FOR THE ROTATION WHICH FOLLOWS.
DRL, 1.2A, 3.25X, -4.88Y, ROTXY-45, PTGRP1

*USE 1" DRILL AT PT 4
ATCHG, TOOL9, GL16, ITD, 70FPM, .016IPR
DRL, PT4, 2.5A

*USE 1 1/4 6 FLUTE END MILL TO FINISH C'BORE.
*LEAVE FINISH BORING STOCK ON OUTSIDE DIAMETER.
ATCHG, TOOL7, GL12.4, TD1.25, 90FPM, .0151PR
MOVE, PT 3, .3X
MOVE, -.475ZB
ICON360, PT3, -.475ZB, 3.18BC, 0CW, TOL.002
MOVE, PT3

*USE 5/8 (4) FLUTE END MILL TO FINISH KEY
ATCHG, TOOL8, GL11.7, TD.625, 90FPM, .0121PR
MOVE, PT 2, 15.4XB
MOVE, -.475ZB
CUT, -3LX
ICON180, PT2, 12.38XB, -.475ZB, 90CW, .7505BC, TOL.01
ICON180, PT2, 15.4XB, -.475ZB, 270CW, .7505BC, TOL.!
CUT, 12.38XB
MOVE, .01Y, 0ZB

*USE 5/16-18 TAP FOR HOLES IN GROUP 3
ATCHG, TOOL10, GL16, TD.31, 65FPM, 18PITCH
FLT3, PT3, 2.625BC, .37RA, .7A, 15CW
FLT, 1A, PTGRP2
BASE, PT13
FLT, 7A, 3.25X, -4.88Y, PTGRP2
FLT, PT14, .7A

*USE 3/8 TWO FLUTE END MILL FOR 3/8 DOWELS
ATCHG, TOOL11, GL12, TD.37, 80FPM, .007 IPR
DRL, 1.1A, PTGRP1

*BASE IS STILL AT PT 13 FOR THE ROTATION WHICH FOLLOWS.
DRL, 1.1A, 3.25X, -4.88Y, ROTXY-45, PTGRP1

*USE 2.125" MICRO BORE BAR FOR HOLE AT PT 3
ATCHG, TOOL12, GL14.5, TD2.12, 200FPM, .0051PR
BORE, PT3, .38RA, 1.82A

*USE 3.187" MICRO BORE BAR FOR C'BORE AT PT 3
ATCHG, TOOL 13, GL15.5, TD3.187, 200FPM, .0051PR
BORE, PT3, .470A
END
```

Milling

```
MACHIN,
IDENT, MDSI MILLING DEMO
SETUP
$ SET BASE 10" ABOVE SURFACE OF TABLE AND 7.5" IN FRONT OF CENTER.
BASE, XB, 10YB, 7.5ZB
$ SET BASE 2" BELOW AND LEFT OF LOWER LEFT HAND CORNER.
BASE, -12XB, -7YB, ZB

$ DEFINE POINTS, LINES, AND CIRCLES ON PART.
DPT21, 2XB, 6YB, ZB
DLN21, 2YB, XB, 0CW
DLN22, 2XB, YB, 90CW
DCIR21, 1EXB, 7YB, 5R
DLN23, PT21, PT(CIR21/CNTR, 5Y)
DLN24, 2+10+5.5XB, 7YB, 47*(32/60)+(27/3600)CW
DCIR22, PT(17.5XB, 2YB, ZB), CIR21, YS
DCIR23, PTLN22/2XL, LN23/2YS), 2R
DLN25, LN21/1YL
DLN06, LN22/1XL
DLN26, LN23/1YS
DLN27, PT(CIR21/CNTR, -.5X), 90CW
DLN31, LN27/1XL
DLN29, PT(CIR21/CNTR), 45CW
DLN30, LN29/.5YS

ATCHG, T00L3, .5TD, 7GL, 900RPM, 91PM
$ MILL PERIPHERY OF PART.
MOVE, 0FFLN21/.25YS, 0FFLN22/.25XS,ZB
CUT, -1ZB
CUT, T0LN22, T0LN01
@CON, CIR23, CW, S(18O), F(PERLN23, YL)
CUT, PARLN23, 0UTCIR21, YL
@CON, CIR21, CW, S(L0C), F(@FFLN24/XL, YS)
CUT, PARLN24, 0UTCIR21, YS
@CON, CIR21, CW, S(L0C), F(TANCIR22)
ICON, CIR22, CCW, S(L0C), F(@FFLN21/YS, XS)
CUT, PASTLN22
MOVE, ZB
$ PERIPHERY COMPLETE.

$ POSITION TO MILL LARGE OPENING.
MOVE, @FFLN25/.25YL, @FFLN28/.25XL
CUT, -1ZB
CUT, T0LN28, T0LN25
CUT, PARLN28, T0LN26
CUT, PARLN26, T0LN27
CUT, T0LN25
CUT, T0LN28
$ MILLING OF LARGE OPENING COMPLETE.

ATCHG, T00L4, .25TD, 1200RPM, 81PM, 6GL
$ POSITION TO MILL SMALL OPENING.
MOVE, T0LN25/.25YL, PASTLN31/.25XL
CUT, -.4ZB, OSTK
CUT, T0LN25
CUT, T0LN31
CUT, PARLN31, T0LN30
CUT, PARLN30, T0LN25
CUT, T0LN31
$ SMALL OPENING COMPLETE.
MOVE, ZB
END
```

Turning

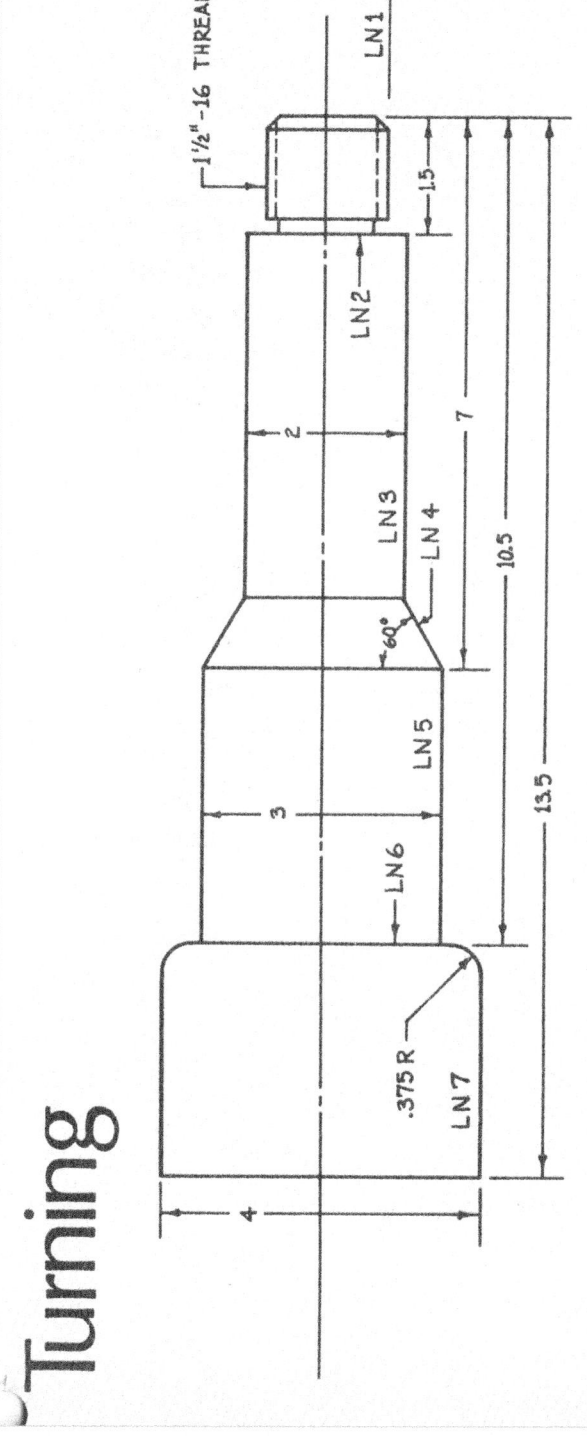

1½" -16 THREAD

LN 1
LN 2
LN 3
LN 4
LN 5
LN 6
LN 7

1.5
2
3
4
7
10.5
13.5
60°
.375 R

MACHIN,
IDENT,MDSI TURNING DEMO
SETUP,16X,20Z

$ SET BASE AT END OF SHAFT.
BASE,17ZB,XB

$ DEFINE LINES ON PART.
DLN1,1.5D
DLN2,-1.5ZB
DLN3,2D
DLN4,-7ZB,3D,30CCW
DLN5,3D
DLN6,-10.5ZB
DLN7,4D

ATCH6,TOOL1,GLX4,GLZ4,250FPM,.02IPR,RANGE2,CON,.03ITLR,.02STK
$ ROUGH SHAFT FROM 4" TO 3" DIA IN 3 PASSES.
TURNX3,MAD,3MID,ZB,-10.5LZ,CR,RAPID

$ ROUGH 3" TO 2" DIA INCLUDING TAPER, CUT NOT TO EXCEED .2 DEEP.
TURNX-,.2SDPTH,MAD3,MID2,ZB,-7LZ,60DE6,CR,RAPID

$ ROUGH 1.5 DIA. IN ONE CUT, DO NOT RETURN TOOL FOR ANOTHER PASS.
CUT,1.5D
CUT,TOLN2
$ THIS COMPLETES THE ROUGHING OPERATION.

ATCH6,TOOL2,GLX4,GLZ4,300FPM,.006IPR,RANGE2,CON,.05TK
CUT,2D,-1.5ZB,-.275LX $ THREAD RELIEF
MOVE,.3X

ATCH6,TOOL3,GLX4,GLZ4,350FPM,.011PR,RANGE2,CON,.03ITLR
MOVE,TOLN1,ZB,CR $ MOVE TO START FINISHING.
CUT,TOLN2
CUT,PASTLN3
CUT,PARLN3,TOLN4
CUT,PARLN4,PASTLN5
CUT,PARLN5,TOLN6
CORNR,.375R,PT(LN6,LN7),SO,F90,CW
CUT,-13.5ZB,CL $ THIS COMPLETES THE FINISHING OPERATION.

ATCH6,TOOL4,GLX4,GLZ4,150FPM,CON,.02IPR
$ CHASE 1 1/2-16 THREAD WITH 2 LAST PASSES OF .001".
THRDX16,MAD1.5,MID1,4D33,2ZB,-1.6LZ,30DE6,.015SDPTH,.003FDPTH,2.001LP
END

COMPACT II Offers Many Unique Features

COMPACT II is a unique, user-oriented, conversational, interactive, time-shared, single-pass N/C processor, capable of operating two, three, and four axis machine tools.

COMPACT II has extensive two axis geometry permitting virtually all point, line, circle, intersection, and tangent computations.

COMPACT II provides the ability to direct the tool relative to points, lines, and circles.

COMPACT II allows the programming of tool diameter and stock allowances.

COMPACT II provides built in canned cycles for drilling, tapping, and boring.

COMPACT II has an easy to use macro capability for special cycles and other repetitive operations.

COMPACT II provides for equally spaced operations on lines, circles, and spirals.

COMPACT II system design and coding provides extremely efficient computer utilization. This concept, normally reserved for small computers was chosen for a large powerful computer so that many more features could be added without penalizing the system efficiency. This is extremely important in time-sharing.

COMPACT II is readily understood by non computer oriented personnel. Based on common English words, it is quickly learned and easily used.

COMPACT II is a free format language thus providing the parts programmer with a natural, versatile and very flexible way to write a given statement.

COMPACT II is essentially machine tool independent.

COMPACT II offers complete editing capability at the time of error discovery and includes the ability for deletions, revisions, insertions of omitted statements, etc., without recompiling. The corrected text is automatically updated during this editing or debugging exercise.

COMPACT II offers interactive operation with on-line

COMPACT II can rotate points in any of the primary planes (XY, XZ, YZ).

COMPACT II has powerful subroutines for enhancing the ratio of output to input for such operations as threading, turning, combinations of facing and turning, taper turning, facing, and boring.

COMPACT II has extensive vocabulary for programming turning machines.

COMPACT II provides machine tool cycle time to assist in production control.

COMPACT II is the only N/C system to provide interactive time-shared N/C programming in a single unified system that does not involve writing intermediate files, and does not overlay program code while the system is running.

COMPACT II combines the speed of a single pass N/C processor with the flexibility and convenience of the post processor concept.

editing thereby significantly reducing the time required to debug a part program. This is a major factor in fast tape turn-around time.

COMPACT II is the only time-shared N/C system to provide tape data transmission verification.

COMPACT II provides the ability for any user to link to our N/C applications engineer when "instantaneous" trouble-shooting assistance is required.

COMPACT II provides either EIA or odd/even parity ASCII tape formats.

COMPACT II's unique building-block design provides an almost unlimited ability to grow. As new or improved features are added, the system is automatically updated so that these new capabilities are available immediately to existing users.

COMPACT II was written by machine tool oriented people fully knowledgeable in N/C machining expertise.

MANUFACTURING DATA SYSTEMS, INC.

P. O. Box 1045
Ann Arbor, Michigan 48106
Telephone (313) 761-7750

An Affiliate of COM-SHARE INCORPORATED

7

The Application Engineer

The next steps in Chuck's plan follow a path that is familiar to any present-day Silicon Valley entrepreneur: he understood that his role as an engineer with a great idea did not automatically make him the best person to bring that idea to market. So when he faced his new responsibility to promote COMPACT, Chuck went back to Bob Guise for one more investment: the funds to hire an application engineer—someone to connect the needs of customers to the programming that Bruce would be doing.

Chuck considered hiring one of the other engineers he knew from Buhr, but he realized that they knew the same things he and Bruce knew. "I needed to bring in new ideas and knowledge." He found that, and much more, in Urbanes "Van" Van Bemden.

■　■　■

Van had become interested in numerical control while studying industrial engineering at LeTourneau Technical Institute in Longview, Texas. As a senior in the fall of 1962 (just as Buhr was setting up its first Sundstrand Model 21 in Ann Arbor), Van had to write a paper on a manufacturing process. When he chose gear manufacturing, his professor pushed him in another direction. Van remembers, "He suggested I write a paper on numerical control (NC), but I had no idea what NC was."

Van jumped into the research, reading *American Machinist*, *Modern Machine Shop*, and the journal of the Society of Manufacturing Engineers; sending away for information from the various companies that were making the NC controls, such as Westinghouse, General Electric, TRW, and Bendix; writing letters to the various machine tool manufacturers, like Cincinnati Milling Machine, Sundstrand, and Giddings & Lewis; and writing to aerospace companies like Boeing and McDonnell Douglas that were using NC machines. The result was a term paper of about fifty pages that Van's new wife, Brenda, typed on an Underwood typewriter during their honeymoon.

As Van prepared to graduate in spring 1963, he inquired with some of the companies from his research and got an interview with Sundstrand. "I flew to Belvedere, Illinois, and got a shop tour, and I was blown away. The NC machines at Sundstrand were the first actual NC machines I'd ever seen. Sundstrand Machine Tool was currently assembling more OM-3 Omnimills (5-axis machines) than existed in the whole wide world. There were machines that were changing their own tools!"

Sundstrand hired Van as an industrial engineer, but just before he showed up for his first day, a position came open in the NC programming department. Van had given a copy of his term paper to Bill Fisher, the supervisor of that department. Bill called to invite Van into his department instead. Says Van, "Now here I was, learning to program 2- 3- 4- and 5-axis NC machine tools using a language called SPLIT."

During this time period, Van remembers, "Every once in a while, a young computer programmer by the name of Chuck Hutchins would show up to check out the progress of a Sundstrand machine that Buhr Machine Tool in Ann Arbor had purchased. He always had to talk to Bill Fisher (my boss) and Hal Baeverstad, who wrote the SPLIT processor, to discuss this and that. I remember that Chuck was very proud of adding the command 'Store Group' (STGRP) to SPLIT, which provided the ability to group a number of points."

■　■　■

Van left Sundstrand in 1966, about the same time Chuck left Buhr. By then, Van and Brenda had two sons, and the family moved to Michigan, where Van got a job with the aviation and aerospace company Lear Siegler in its instrumentation division in Grand Rapids. After about a year, he transferred to that company's new fabrication technology division in Zeeland, Michigan. That plant had several different kinds of NC machines, including Kearney & Trecker's Milwaukee-Matic Model 2 with shuttle capabilities and a half-dozen Milwaukee-Matic Ebs. Says Van, "We were using IBM's Autospot, a 2.5-axis NC programing language, but the computer was in Grand Rapids, some thirty miles away. Thus the turnaround time to correct programming changes and mistakes was not very good. We were looking for a solution to this problem."

In early 1968, Van attended a meeting of the Western Michigan Chapter of the Numerical Control Society, where he saw Chuck Hutchins—that guy who used to visit Sundstrand to talk about SPLIT— demo his newly created NC programing language called COMPACT. Van recognized it as a "SPLIT knock-off," but the key advancement was that this demo included off-site computer timesharing.

After the meeting, Van arranged for Chuck to come to the Zeeland plant to give his demo to the Lear Siegler team. "A few days later," Van remembers, "Chuck, with his trusty ASR33 teletype, came out to show us what he had and how it could solve our turnaround problem. It was really funny, because I had written the program that we ran through his COMPACT NC program. After processing it via Comshare's timesharing SDS 940, we actually ran it on one of the Milwaukee-Matic Eb machines. Everything worked fine." (The Kearney & Trecker Milwaukee-Matic Eb was one of the first machine tool links that Chuck and Bruce wrote.)

After some discussion about functionality and cost, Van and Chuck went to lunch together. Van confided to Chuck that he was thinking seriously of going back to Sundstrand to work in their sales department doing demos. "Don't do that!" Chuck said, recognizing the serendipity of circumstances. He had just gotten permission to add that application engineer to his little Comshare team. "Come to work for me." It was a turning point in Chuck's commitment, because hiring Van meant he had taken on the financial responsibility for a family with two kids.

Van talked to Brenda that evening about the opportunity, and by the following weekend, he was visiting Ann Arbor. When Van arrived at Chuck's house, he found Chuck in his jockey shorts, sitting at an ASR33, communicating with a computer jock in Hackensack, New Jersey, where Comshare had one of its SDS 940 computers. After working in big, established corporations, Van was startled by the informality of this

Urbanes "Van" Van Bemden with his wife, Brenda, and sons, Jeff (left) and Gary, 1969. Photo courtesy of Urbanes Van Bemden.

start-up operation. "I thought, *Holy smokes, what am I getting into?*" Chuck dressed and took Van to Comshare's headquarters in Huron Towers to see the SDS 940 and the office where Van would work.

Urbanes "Van" Van Bemden joined Comshare on June 16, 1968. It was a Sunday, but Van and Chuck worked all day to prepare for a series of sales calls that Chuck had lined up.

Early the next day, they were on the road to Traverse City. "The demos went long into the evening both Monday and Tuesday," Chuck remembers, "and as we were driving to yet another demo on Wednesday at noon, Van looked over at me and exclaimed, 'Chuck, do you realize we've put in a forty-hour workweek already?'" Such was the schedule for a start-up.

The demos went well, though Chuck and his team weren't quite ready to sign up customers. They were still trying to get a feel for what COMPACT could do, how to customize it to customers' needs, how to price those services, and whether this could really be a viable business.

HOT TECH COLD STEEL

Back at the office, Van began writing the first *COMPACT User's Manual* to replace the only other documentation so far produced, the original COMPACT brochure. He also began to write programs himself, and he uncovered some untried features that weren't working and needed to be fixed.

■ ■ ■

By this point, Chuck had been living off the budget of Comshare for more than a year without bringing in any income, and now he'd added two more employees. The time had come to start selling COMPACT, but for that, Chuck knew he needed business management expertise.

He thought this need would be met through Comshare, but Bob Guise already had his hands over-full with other priorities. The company was running out of money, and in July 1968, Comshare filed an IPO with the Securities and Exchange Commission to take the company public. That process required an enormous amount of Guise's time, on top of the constant anxiety about funding.

Guise still wasn't paying much attention to Chuck's endeavor: "I knew he had hired Bruce Nourse and Urbanes Van Bemden, but I'm ashamed now to admit that I didn't realize how good these people were." As Chuck got closer to launching his product, Bob felt that Comshare's existing commitments did not allow for the salespeople needed to sell COMPACT. "And I had learned the hard way that whatever a product costs to produce, multiply that number by ten to predict what it will cost to bring it to market."

Bob realized he was going to have to spin the NC business out of Comshare. He consulted with the Comshare board, and everyone agreed. Guise also consulted with the investment banking firm of Ball Burge & Krause in Cleveland, where he had developed a friendship with Comshare investors Harry Altman and Ed Paran. Guise remembers, "Over the course of two meetings, we agreed on four points: this was not a BB&K kind of financial deal, we needed a private investment banker, we needed a seasoned executive officer to lead the new company, and I was already spread too thin to do that job."

In early July of 1968, Bob Guise called Chuck into his office. Before Chuck had even sat down, Bob said, "Chuck, I'm running out of money, and I think you should start your own company."

As the words sunk in, Chuck felt instant panic. He had trouble controlling his emotions as he replied, "Bob, I know a little about NC machine tools and a little about computers, but I don't know anything about running a business. If I try to do this myself, the wolves of the business world will eat me alive!" With that, Chuck stormed out of the office.

A few days later, having cooled off, he returned to Bob's office and said, "If we can find someone who understands as much about business, marketing, finance, and management as I understand about computers and NC machine tools, I think that starting a new company would be an exciting thing to do. But without such a person, it would be a disaster."

Bob agreed. He wanted to see Chuck and COMPACT succeed, and he wanted Comshare to help make that happen—especially because he expected Chuck's business to become a customer for Comshare. "As I was pondering the problem," Bob remembers, "a faculty friend in the Michigan Business School called. He knew of a sharp individual who was leaving Bendix Corp., right here in Ann Arbor, and he wondered if I would be interested in talking to this man about a possible position at Comshare."

In September 1968, Bob Guise met Ken Stephanz for lunch on North Campus. Guise remembers, "The lunch was pleasant, but there was a certain tension in the air. An old expression of my dad's kept running through my mind: 'Two women kissing is like two boxers shaking hands.' The lunch ended about two o'clock with both of us politely agreeing that there was no suitable match at Comshare."

But as Bob was walking to his car, he had a nagging feeling that he was letting a good man get away. He knew that Ken had the right qualities to do great things. Ken was about to get into his car when Bob called back to him, "Wait a minute! Have you ever heard of an animal called numerical control?"

Ken looked up and grinned. "Of course I have. I just came from Bendix."

Bob laughed. "I should have thought of that." Bendix had been making controllers for NC machines. Bob said, "Would you like to come back to my office and talk about another opportunity?"

8

The Chief Executive

Ken Stephanz was forty-one when he met with Bob Guise to talk about numerically controlled machine tools. He had been in Ann Arbor for three years as an executive with Bendix Corporation to start that company's "electro-optics division." It was a position that drew on his previous career experience in developing manufacturing plants, managing large teams, negotiating government contracts, and engineering precision instruments. Before Bendix, Ken had also started his own high-precision metalworking company, served as chief engineer and then plant manager at a Fortune 100 company, and saved two other businesses from bankruptcy. Add to that a basic knowledge of numerical control, and Ken seemed almost tailor-made for the challenges of launching COMPACT into a profitable business. His career had always been on a fast track, and it was about to shift into overdrive.

■ ■ ■

A child of the Great Depression, Ken started earning his own way at age twelve. Kenneth and his fraternal twin brother, Harold, were born in Toledo, Ohio, on December 19, 1926, to Francis and Mary Edith Stephanz, whose first son, Glenn, was just a year older than the twins. Francis worked as an engineman for the Pennsylvania Railroad, but he was laid off soon after the 1929 stock market crash. About the same time, Francis and Edith discovered that the mortgage payments they'd been making through a real estate agent had not been submitted to the bank; their house was in foreclosure.

In search of a fresh start, the Stephanz family drove to Grand Island, New York, a large, rural island in the middle of the Niagara River, to visit Edith's sister's farm. After a two-week stay, Ken's parents decided to move there, so they returned to Toledo to pack up their belongings.

But in their absence, their house had been emptied of everything they'd owned. With nothing but the clothes they had packed, they turned around and headed back to Grand Island.

"My folks started the Depression with three small children and two suitcases," says Ken. "That was it." This hardship forced Ken to grow up fast as he learned to work hard, solve problems, and fend for himself. He still takes pride in these lessons today as a ninety-four-year-old widower living on his own in South Florida.

After two years on the family farm, Ken's father got a job with the Works Progress Administration—one of President Roosevelt's New Deal programs—and the family moved into a summer cottage on the east side of Grand Island. They lived there through two brutal winters with only a wood-burning potbellied stove for heat. "We lived at the eastern end of a 200-mile wind tunnel called Lake Erie," Ken jokes. He started school at a one-room schoolhouse.

When the railroads were operational again, Francis was hired back, and the family moved to Sandusky, Ohio. This is where Ken spent the remainder of his childhood, but the family changed houses frequently. "In my first seven grades, I went to seven different schools. You quickly get to be an outsider looking in." Ken feels that these early experiences helped shape his "lone wolf" attitude toward life. "I don't know if I started that way, but that's what I turned out to be."

Ken's first job, at age twelve, was for a dressmaker, delivering packages on his roller skates for ten cents a trip anywhere in town. Soon after, he worked as a busboy for a high-end restaurant, then as a waiter, and then as a fourteen-year-old bartender (when all the older male employees were drafted into World War II). As war production ramped up in the industrial Midwest, Ken got a job at a steel foundry that was making parts for warships. During his last two years of high school—after taking a couple of drafting classes—he worked afternoons and evenings as a design draftsman in a defense plant. He still managed to graduate third in his high school class of 300 and receive the Bausch & Lomb National Honorary Science Award plus two college scholarships.

While still a senior in high school, Ken lied about his age and enlisted in the U.S. Naval Aviation program to become a pilot. Two days after graduation in 1944 (when he was still just seventeen), he went on active duty. He eventually ended up at Alameda Naval Air Station where he decommissioned ships returning from the Pacific Theatre as the war came to an end. He was fascinated by the planes of all sorts that flew in and out of the Air Station; it was an interest that would stay with him as he returned to civilian life.

During his navy years, Ken took math and physics classes through Oberlin College and the University of Wisconsin, and he headed back to college as soon as his military service ended. At Ohio State University, he enjoyed science more than math and thought he wanted to be a chemical engineer, "but I got so bored memorizing chemical formulae that I switched over to electrical engineering. And that was particularly good for me."

■ ■ ■

After graduating with honors in 1948, Ken spent the next twelve years of his career at International Telephone and Telegraph (ITT), headquartered in New Jersey. He started as an engineer but quickly proved himself an able manager and, at age twenty-three, became chief engineer of its 150-employee engineering department.

While living in New Jersey, he also auditioned for and earned a spot in the famed New York City–based Robert Shaw Chorale. Ken, whose baritone voice has a wide range, had studied vocal performance in college, worked as a paid church soloist, and took voice lessons from an opera teacher in New York. He sometimes wondered what his life would have been like if he had pursued a music career full time, and singing with the professional Chorale was a close approximation. The hours were long, with travel and recording requirements, but the performances were a heady experience of shared excellence. "It was a facet of my life that I just really loved," says Ken.

Eventually, ITT sent Ken (along with his wife and two children) to Roanoke, Virginia, where he set up a new engineering and manufacturing plant to build government-classified electronics equipment. "I ended up building a brand-new plant there." He was just thirty years old when he took over management of the plant and its 200 employees.

Never having taken formal management classes, Ken says he flew by the seat of his pants, but his first rule was "The answer is yes." Whatever his bosses asked for, he said, "Yes, I can do it," and, yes, he did do it. He had high expectations for himself, which translated into high expectations for everyone who worked with him. "I'm sure I pissed off a lot of people," Ken says of those early years in a leadership role. "If you did your job, you were really happy working for me; if you didn't do your job, you were really unhappy working for me. I expected results." And he got them. "We made so much money off that government contract, that the government forced us to renegotiate the price and give them some money back."

Ken Stephanz at age thirty-three, during his time managing an engineering and manufacturing plant for International Telephone and Telegraph (ITT), 1959. Photo courtesy of Ken Stephanz.

After a few years in Roanoke, Ken left ITT to work for an electronics company in Boston, and then returned to Roanoke to start his own high-precision metalworking machine shop, Microproducts, Inc. With start-up funding from other local leaders, he secured government contracts to build specialized parts for nuclear submarines.

"I did all the selling myself," says Ken. "But my favorite part was managing people. I liked hearing their problems and helping them find solutions. I liked helping them grow from wherever they were to something better. I liked getting them to the point where they could do better than they even thought they could. That's a wonderful experience."

Ken learned some tricks over the years for motivating employees. He remembers one young factory worker—we'll call him "Joe"—

who was skilled and smart but not much inclined to do his job. The manufacturing manager, Frank, was at the end of his patience, so Ken called them both into his office. "I said to the young man, 'Joe, I'd like to fire your ass. You don't do this; you don't do that. You screwed this up. You screw up everything. I'd like to fire you right now. But Frank here thinks you're great. So here's what we're going to do. You're going to work exclusively for Frank now, but if you screw up even once, he's going to fire you. Now, get the hell out of my office.' It worked like a charm. Joe became one of our best employees."

Ken valued the control he had as owner of the company. "I couldn't wait to get up in the morning. I couldn't wait to go to work. I didn't want to leave in the evening." Microproducts, Inc., had about fifty employees when Ken sold the company in the early 1960s during a painful divorce.

When he moved to Buffalo, New York, for a job as a turnaround expert, he found himself back on the shores of Lake Erie, this time as CEO of Forbes and Wagner, an electronics assembly company located in Silver Creek, New York. With about 300 employees, the company was a major employer in the small village, and it was about to go bankrupt. "Their losses were at least 30 percent of their sales," Ken remembers.

He immediately called a meeting of the company's creditors. "I treated them like a board of directors. I said, 'Give me some time, and I'll turn this company around.' Five months later, we were in positive cash flow." In addition to improving production, Ken renegotiated contracts with the company's customers, which included the U.S. Air Force and big corporations like General Electric. "The advantage of a turnaround is that any decision you make is probably going to be better than what was done before. The important thing is to do something every day. If you need to go back and re-do it, that's fine, but you've got to make decisions quickly and move on."

One of Ken's decisions was to limit the one-hour lunch break for senior staff to thirty minutes, just like the factory workers. "The senior staff were so depressed," he remembers, "they were going out every day, taking at least an hour and a half, and drinking their lunch." The shortened lunch break forced everyone to eat together in the company canteen—and remain sober. Before they knew it, "productivity shot up dramatically."

But all these measures didn't prevent the need to lay off some employees. "It's the toughest management decision there is," says Ken. "It never gets easier. Every single person was extremely difficult to let go. And did I procrastinate? Yes. So does everybody when making that decision."

Word soon spread of Ken's success with Forbes and Wagner, and corporate headhunters put him on their lists. Before long, Bendix came calling.

■　■　■

In the 1960s, Bendix Corporation was about as diversified as an American company could be. It was started by inventor Vincent Bendix in Chicago in 1914 and was officially formed in South Bend, Indiana, in 1924. Although its original purpose was to build brake systems for automotive manufacturers, Bendix soon expanded into the field of aviation, making hydraulic and electronic systems for aircraft. Contracts with the government during World War II led to an enduring connection between Bendix and the military. Among other projects were the design and manufacture of missiles, torpedo electronics, radio transmitters, radar equipment, and telemetry systems.

In the 1950s, the company added consumer electronics to its portfolio, with car radios, phonographs, and televisions. Starting in 1956, Bendix also produced an early computer, the Bendix G-15, in collaboration with UC–Berkeley professor Harry Huskey. Following on this toe-dip into the computer industry, Bendix also began building controllers for numerically controlled machine tools.

As it diversified, Bendix built new plants in the Midwest, New Jersey, and California. In Ann Arbor, the Bendix systems division included an "automatic mass transportation systems group" as well as the Bendix aerospace division. By 1958, most of Bendix's Ann Arbor operations were located on a 43-acre campus on the corner of Plymouth and Green Roads (near U-M's North Campus). This facility engineered and built major weapons systems with technological advancements in guidance and control, infrared, radiation resistance, aerodynamics and propulsion, radar, and acoustics for military strategies such as global weather reconnaissance, underwater surveillance, and supersonic aerial target systems. By the early 1960s, the Ann Arbor division was developing satellite radio relay systems.[10]

Ken was recruited in 1965 for Bendix's newest endeavor in Ann Arbor: the electro-optics division, which would make night-vision devices for soldiers in Vietnam. The classified program was dubbed "Taking the Night Out of Charlie." A new plant was needed, and Ken was hired to design and oversee its construction, and then staff and manage it. "It was a very difficult scientific project," he remembers. "The night vision had to work when it was totally dark, no moonlight, no ambient light of any sort. This device took the darkness and made it into light by roughly several million to one." The technology to do this came out of the Bendix research laboratory, but Ken's team put it into practice. "And we were highly successful. We started with just me on the payroll, and after the plant was built, we grew to about two hundred employees in three years."

By then, Bendix's original Ann Arbor facility had begun its most celebrated accomplishment: designing and building the Apollo Lunar Surface Experiments Package for the 1969 lunar landing. This "package" was essentially two suitcase-sized pieces of equipment—a seismometer and a laser-ranging retro-reflector (to measure the moon's orbit and distances between the earth and moon)—that would be set up by the astronauts and left behind on the moon's surface. These machines worked exactly as intended for nearly a decade afterward.

One other relevant tidbit from the history of Bendix in Ann Arbor is its 1969 acquisition of Buhr Machine Tool Company. By that time, of course, Chuck Hutchins and Bruce Nourse were gone from Buhr. And Ken was gone from Bendix.

■ ■ ■

Ken resigned from Bendix when his job to build and staff the electro-optics plant was completed. But he had come to enjoy life in Ann Arbor, with its Big Ten university and attendant attributes—

music, arts, cultural events, and sports. He had also met Edie (who would become his second wife), who lived in the area.

In addition, Ken had renewed his interest in flying since his World War II days and had begun taking flying lessons out of the Ann Arbor airport. Ken eventually became licensed to fly turboprops and small jets and to fly on instruments. (He achieved about 6,000 pilot hours as "pilot-in-command.") About instrument flying—in the dark and/or in clouds, with nothing to see—Ken says, "It's the most relaxing exercise I've ever done. You have to totally focus, or you are gone, so I focused to the extent that I didn't think about anything else. I found that very relaxing."

But he wasn't ready to relax in his career, and when Bob Guise told him about COMPACT, Ken was intrigued. He asked Guise to cover his expenses for a week or two to size up the situation.

Bob remembers, "It was a no brainer! I was convinced Ken had considerable ability, and that Comshare had everything to gain and nothing to lose. If Ken turned down the NC idea, his reasons would be valuable information." However, if he came back ready to give Chuck's ideas a chance, Guise would be able to spin COMPACT out of Comshare and on to potential success.

Ken spent about ten days traveling to manufacturing companies that were using NC machines, so he could see the technology at work. He also visited Cincinnati Milacron and other machine tool builders, seeking to educate himself and explore the market possibilities. He remembers, "I went to as many different machine tool manufacturers as I could think of, and then I sat down and wrote a 'dissertation' about numerical control, just to be sure I understood it." He learned about APT and other competing languages and could see that they were "ponderous." There was clearly a market opportunity for something better.

When he reported back to Bob Guise that he wanted to explore this opportunity further, Bob set up a meeting for Ken with Chuck Hutchins. "November 11, 1968, counts as one of the most significant days of my life," Chuck remembers.

The three men met up at Weber's, the well-known, family-owned hotel/restaurant on the west side of Ann Arbor. Bob Guise made introductions, bought a round of drinks, and soon excused himself. Chuck and Ken ordered dinner and talked. And talked and talked. At 11:00 p.m., closing time, the manager politely asked them to leave.

That was the night Ken began to think of Chuck as a "wild 'messianic' guy who believed he could change the numerical control machine tool manufacturing world single-handedly." But Ken also concluded that, "Whatever Chuck was, he was not stupid." As he drove home in the dark, Ken had to admit, *That was the most stimulating, invigorating, interesting, and arresting discussion I've had in a very long time.*

Chuck came away trusting that Ken could, in fact, be the CEO he needed to launch a company. With Bob's blessing (and Comshare funds), the two men spent the rest of 1968 going on demonstrations together as they continued to field-test COMPACT, research the competition, and explore the market potential.

One such demonstration took place at a company in Chicago. Ken, eager to use his pilot license, had the idea to fly to Chicago in a single-engine, two-seat Piper Cherokee Arrow. Ken asked Chuck if he wanted to fly with him. "He gave me a choice, not forcing the issue," Chuck remembers. "I agreed, though it would be my first flight in a small plane. Our destination was Meigs Field on the Chicago waterfront. It was early December, and the wind was blowing plenty hard from the west, making for a tough cross-wind landing. I watched as Ken came down, crabbed at a pretty good angle. I looked down at the cold water of Lake Michigan, and then I noticed a DC-3 with its nose in the chain link fence, the left prop blade bent by a collision with the fence." Within seconds, Ken set the Piper down gently without mishap, and Chuck breathed a sigh of relief.

At each demonstration, Ken could see that COMPACT was well received. He even practiced programming the machines himself, so he could get a feel for how easy it was to learn. That's when he asked for the head-to-head comparative test with a General Electric timesharing customer in Cleveland. The test opened Ken's eyes to how much they could charge and still beat the cost of the competition. Says Ken, "We were trying to confirm Chuck's claims that his technology was faster, easier, better, cheaper. Indeed, it *was!* I mean, we felt like Jesus Christ's brother. By the time we got through with our demonstration, people were saying, 'When can we get this?' It was incredible."

Ken immediately began working on a detailed cash flow projection—writing it by hand on ledger paper, of course, since this was long before the invention of Microsoft Excel. "I painfully struggled through it," says Ken. "It took hours and hours and hours to put together." He had to make a seemingly endless series of assumptions: How many sales calls can a salesman make a week? How many times will he have to go back before he makes a sale? How many of those total sales calls will turn into actual sales? Ken made educated guesses about customer growth, staffing needs, computer sharing costs, travel costs, and so forth. The resulting thirty-line, thirty-six-month spreadsheet showed the new company breaking even in the twenty-second month of operation. But after that, profits would increase quickly and steadily. Who would keep them afloat until then?

9

The Founding

It was a cold winter day in Cleveland when Ken and Chuck pitched their business idea to venture capitalist David Morgenthaler. The meeting took place on a Saturday morning in December 1968 or January 1969. They met at Dave's office at 9:00 a.m., talked with him until noon, had lunch together at Dave's private club, and continued talking into the afternoon. Around 3:00 p.m., as Ken and Chuck prepared to leave, Ken asked Dave directly, "What do you think?"

Dave smiled and said, "Ken, there was a guy here yesterday to pitch me a business idea at 9:00 a.m. He was gone by 9:05."

When Ken and Chuck were safely out of earshot, Ken said, "I think we just hooked our first big fish!"

■　■　■

David Morgenthaler was indeed interested in the business that Ken and Chuck proposed. He had good reason to be, and not just because of Chuck's solid product and Ken's "wild visions" of future profits. First, Dave knew Bob Guise and knew that Comshare was committed to the success of the effort. It was Bob who sent Ken and Chuck to talk with Dave.

Second, Dave already knew Ken Stephanz. In fact, Dave had interviewed Ken for a job when Ken was still at Bendix. Ken turned down the job offer—it just wasn't something that interested him—but the men had impressed each other. "Dave and I had quickly developed an excellent relationship," Ken remembers.

Third, this business was well suited to Dave Morgenthaler's background. He had a bachelor's and a master's degree in mechanical engineering from MIT, and after serving in World War II, he had worked in the manufacturing field. He headed up the sales division for the world's largest manufacturer of jet-engine fuel nozzles, and then he ran the Cleveland division of Foseco, which manufactured products for foundries and steel mills. As a fast-track executive, he joined the Young Presidents' Organization, eventually becoming YPO's international senior vice president.

Morgenthaler was recruited to Foseco by J.H. Whitney & Company, one of America's first venture capital firms, formed by the Whitney family after World War II to manage investment requests from entrepreneurs. As Morgenthaler engaged with the venture capitalists on his corporate board, he began to make a few investments of his own in start-up companies in the Cleveland area.

When he sold his interests in Foseco in 1968, Dave used his own capital to found Morgenthaler Associates. Just a few months later, Chuck and Ken came calling.

At that time, few venture capital firms existed. Dave would go on to be a national leader in the industry, helping to found the National Venture Capital Association. He would eventually fund more than 325 start-up companies in information technology, life sciences, and industrial technology, including Apple, VeriFone, Evernote, and Siri.

David Morgenthaler (left) and Bob Pavey of Morgenthaler Associates, the Cleveland-based venture capital firm that took a chance on MDSI. Photos courtesy of Gary Morgenthaler.

HOT TECH COLD STEEL

But in 1968, at age forty-nine, he was just getting started, and he was looking for a great opportunity.

He was also looking for an able assistant, and he found one in Bob Pavey, a newly minted MBA from the Harvard Business School who had studied physics at William & Mary, and metallurgical engineering at Columbia University. Dave Morgenthaler had hired Pavey, age twenty-five, to work at Foseco, where Pavey became responsible for new technology involved in continuous casting. He left Foseco and joined Morgenthaler Associates in the fall of 1969, just as it was becoming clear that Chuck and Ken's venture "had the potential to be a big success," as Bob remembers it. "So Dave assigned me to that investment. I was 'the young kid.' It was an amazing opportunity."

Two days after Chuck and Ken had been in Cleveland, Dave called Ken to say he would back them. Soon after that, Dave came to Ann Arbor to hammer out a deal. Ken, Chuck, and Bob Guise met Dave at his hotel. Guise remembers, "We roughed out an investment deal on the back of a large manila envelope while sitting on a hotel bed."

The plan was for Morgenthaler Associates and Comshare to each invest $300,000 in three $100,000 installments, for an initial total investment of $600,000. To cover his half, Dave invested $200,000 of his own capital and $100,000 from a group of Cleveland businessmen who were looking for investment opportunities.

Morgenthaler was named chairman of the new company's board of directors, which also included Bob Guise and Ken Stephanz. (Bob Pavey would join the board later that year.) Chuck and Ken were considered the company founders, with Ken as president and CEO, and Chuck as vice president for research and development. Bruce (as the chief programmer for research and development) and Van (as the first application engineer) were founding employees.

Van recalls how the company name was chosen. "We tossed a few names around and settled on Manufacturing Data Systems, which would be abbreviated as MDS. However, that abbreviation was already in use by someone else, so Ken suggested we incorporate and call it MDSI." In addition, the team changed the original full name of the software from "Comshare's Program for Automatically Controlling

Tools" to "Computer Program for Automatically Controlling Tools," at which point it was just called COMPACT.

Manufacturing Data Systems, Inc. officially spun out of Comshare on Saturday, February 1, 1969. The first day of business was Monday, February 3, 1969, Chuck's thirty-fifth birthday. As Chuck still likes to say, "Life really does begin at thirty-five!"

■ ■ ■

The team of four set up their first offices at 2223 Packard Road, south of Stadium Boulevard. The wastewater engineering firm of McNamee, Porter and Seeley had just built a two-story addition onto their building, and they rented the second floor and the basement to MDSI. Ken and the sales department would be on the second floor, with R&D in the basement. Chuck had a second-floor office, but he spent most of his time in the basement with his software engineers. He remembers, "It was quiet down there, and no one bothered us."

As Chuck remembers it, Ken was all about office protocol; he laid out office space and purchased office furniture according to strict hierarchical rules. The CEO's office was biggest, the VP's office was a bit smaller, and the others shared a two-person office. The CEO's desk had overhang on three sides, the VP had overhang on one side, and all other desks had no overhang. The CEO had a high-back chair, the VP had a chair with arms, and the others had chairs without arms.

As all this was unfolding, Chuck saw a cartoon in *Better Homes and Gardens* magazine that showed King Arthur speaking to the Knights of the Round Table, saying, "The significance of the round table is that we are all equal. The significance of my high-back chair is that I'm just a smack more equal." Chuck copied the cartoon and placed it on Ken's desk early one morning. Ken never acknowledged it, so Chuck put another copy on his desk a few days later. "I nagged him a little," says Chuck.

It was 1969, after all, and the buttoned-up corporate culture was beginning to relax. But Ken was old-school. He still wore a suit and tie to work every day. Chuck had worn a tie since he was in high school at

Cranbrook and had switched to bow ties, which were safer around the machine tools in the school shop. When he went to work for Buhr he continued the practice. But the software guys in the basement typically dressed more casually. At one point, Chuck stopped shaving, thinking he might grow a beard. Ken said nothing until a Friday, when he and Chuck were planning a trip to Caterpillar the following Monday.

"Are you going to Caterpillar with your beard?" Ken asked.

Chuck only hesitated for a moment. "No, sir," he said, and shaved that weekend.

Chuck remembers, "I had so much to do just getting my team going that I had to trust Ken's leadership. Besides, I didn't have much choice. He was the CEO. I didn't want to mess up."

Plus, Chuck was grateful for Ken's business and managerial skills. He began to think of Ken as "Manager Cum Laude." Chuck remembers, "I always liked his method of conducting staff meetings. He constantly kept his ear tuned to problems within the organization, and at each meeting, he would bring up what he saw as a problem. Then he would let everyone speak his piece about it. After the once-around, he would ask if anyone had anything else to say. After hearing second thoughts, he would say, 'This is what I think we should do.' Seldom was there any disagreement. Everyone felt involved in the decision making."

Ken confirms that these discussions were essential to his own decision making. He says he reversed course many times after hearing what the employees thought of a problem. "I often changed what I thought I was going to do, based on their feedback."

Staff meetings quickly grew in size as MDSI began hiring necessary additions, including Roy Winn, who came from Bendix to work in sales, and the company's first secretaries, Joan Matthews and Elizabeth "Betty" Ruddy. Ken lured Betty away from Comshare to be his executive secretary. Married with five children, Betty was in her forties when she joined MDSI. Her maturity, no-nonsense attitude, and operational skills were invaluable to the team, and she assumed multiple informal roles, from office manager to gatekeeper.

From the start, MDSI followed a convention of numbering employees sequentially, creating a list that would quickly expand. Ken was

Betty Ruddy, MDSI office manager, in a light moment. Photo: Bruce Nourse.

considered employee #101, Chuck #102, Bruce #103, and Van #104. Roy Winn then became #105, and so on.[viii] (Chuck liked to tease Ken that "rank has its privileges" since Chuck really was the first member of the company, and in fact, Bruce and Van preceded Ken as well.) An employee's number was a clear marker of how long he or she had been with the company, and from then on, employees would know their own numbers and compare with others.

Though everyone was hired for a specific role, Van remembers that "Back then, it didn't matter what your job title was, everyone wore ten hats. We simply did what needed to be done." As an application engineer, Van was doing demos to potential customers and working on a software manual, but he also trained new employees and customers in how to use COMPACT II.

■ ■ ■

As more staff came on board and the budget tightened further, Ken and Chuck jumped at the opportunity to meet with another potential investor. In the spring of 1969, they traveled to New York City to meet with Sanford Weill who was, at that time, the head of a prominent securities brokerage firm. Ken and Chuck took a few minutes to explain

viii John Morris was #106, Joan Matthews #107, and Betty Ruddy #112.

HOT TECH COLD STEEL

to Weill about MDSI and the use of the SDS 940 to program NC machine tools using computer timesharing.

Weill understood the concept because he knew Tom O'Rourke, the president of Tymshare in Cupertino, California. Recall that Tymshare was Comshare's one-time collaborator in developing the SDS 940 and the computer timesharing concept. Around the same date that Ken and Chuck met with Weill, Xerox had acquired Scientific Data Systems, the originator of the SDS 940, and the computer was forever after known as the XDS 940.

With Chuck and Ken still sitting in his office, Weill called O'Rourke for a reality check. "Hello, Tom," he said. "I've got two gentlemen here in my office who say they are going to use an XDS 940 to program NC machine tools."

Apparently, O'Rourke's reply was one word: "Impossible."

Weill thanked Tom, hung up, thanked Chuck and Ken for coming, and bid them adieu. Chuck did not appreciate Weill's (or O'Rourke's) dismissive attitude, but Chuck would soon have the last laugh: "Within a couple of years, we were one of Tymshare's largest customers, using five dedicated XDS 940s to program our customers' NC machine tools."

Chuck is too respectful to point out how foolish Weill was to turn down an opportunity to be an early investor in MDSI. Not that it curtailed Weill's good fortune. He would become known for his role as CEO of Citigroup. He is also a significant donor to the University of Michigan; the Gerald R. Ford School of Public Policy is now housed in Joan and Sanford Weill Hall.

■　■　■

Even as Ken Stephanz was running the start-up company, looking for additional investors, preparing financial reports for the board, and buying office furniture, he also served as MDSI's lead salesman. Following up on Chuck's earliest demonstrations, Ken made all the initial sales calls, sometimes with Chuck, sometimes with Van, and soon with others as they were hired. Ken remembers, "I'd go out with a new salesman, and after every sales call, we'd sit down and have a face-to-face about what we should have done or shouldn't have done." The key to good sales was listening to the problems of the potential customer and then adapting the MDSI message to that company's needs. "All you had to do was listen," says Ken.

One of the first customers to sign on was Jack Clausnitzer, a former colleague of Chuck's from Buhr, who had opened his own machine shop, Brighton NC, in Brighton, Michigan. Jack's interest in COMPACT was primarily the savings in turnaround time for producing the punched paper tapes he needed for his NC lathe. Before he signed on with MDSI in April 1969, Jack used a computer located twenty miles from his shop that was available only at night.[11]

Another early customer was the GM Tech Center in Warren, Michigan. Van went there to demonstrate how COMPACT II could program a cylinder head, but he quickly discovered that he needed the ability to rotate the cylinder head on a trunnion. At that time, COMPACT II only had the ability to rotate a point in the XY-plane (with the command "ROTXY"), but a successful program for the trunnion fixtured part would need to give instructions for the YZ-plane (ROTYZ) and the ZX-plane (ROTZX). Van remembers, "I called Ann Arbor from the Tech Center and talked to Bruce and Chuck. Overnight, they added that ability, and I went back to GM the next day to show them we could do the job. The GM Tech Center went on to become one of many GM plants to use our system."

When Van did a demo on July 20, 1969, at a General Electric plant in Cincinnati, his contact was a former colleague from Van's time at Sundstrand, Harold Anderson. "At GE, Harold had the same problem that I had faced working at Lear Siegler," says Van. "He had no ready access to a computer to program the NC machines. Since COMPACT II was based on Sundstrand's SPLIT, Harold had no problem following me as I programmed his part, and I felt pretty good about his becoming a customer." Harold then invited Van to his house for dinner, where they watched the news of the Apollo moon landing on television.

Van was also called in when another new customer, Claude Wilson of Wilson Concepts in Dayton, Ohio, needed urgent help in programming his NC machines. It was a Saturday morning, but Chuck had agreed to meet Mr. Wilson at the MDSI office, and then he called Van at home for help. Van remembers, "Wilson had some aircraft parts (gear boxes, etc.) that he needed, like, yesterday! I spent the rest of the day programming some of his parts. What I didn't know was that, while I was programming, Claude asked Chuck if I could come down to his plant and help get the programs going on his machines. So later that Saturday, I was on my way to Dayton."

Wilson had several Lodge and Shipley NC lathes, and MDSI had just written the machine tool link for that particular model. Van used that software (and an ASR33 teletype machine that he'd brought along from Ann Arbor) to program several parts for Wilson. He remembers, "We loaded the tape into the machine, but no matter how hard we tried, we couldn't get a good part off. We put dial indicators on the axis and added up all of the X and Z movements, and they seemed to add up correctly. After checking my program over and over, I finally found the problem: Bruce had forgotten to reinstate the G01 after a G04 dwell cycle." Van called Bruce, explained what needed to be fixed, and within thirty minutes, everything was running smoothly. Says Van, "That's what I call customer service at its best."

Claude Wilson was a loyal, longtime MDSI customer and turned out to be an especially cool character. Van noticed that Claude always carried a saxophone in his car and wasn't shy about sitting in with a band at area nightclubs. "I witnessed this several times," says Van. "What I didn't know then, but found out years later, was that Claude Wilson had been first chair saxophone with the great Jimmy Dorsey Orchestra."[12]

■　■　■

Every day held the potential for signing new customers and also the potential for customers to request refinements to COMPACT II to meet their unique needs. R&D had to ramp up its programming for each new request. Chuck remembers, "I was working my heart out just to keep my guys going." To ease the burden, Chuck hired more engineers for his team, including Don Willan (#117) and Don Colley (#120), who both turned up in late 1969.

Though Bruce had seniority within the growing R&D team, he was never asked to assume a managerial role in the company; he was too valuable as a programmer. Bruce explains, "I am what's called an individual contributor. I work best if we come up with something we're going to do, and then it gets handed to me, and I have free rein to figure it out. I very often go through two or three software architectures before I settle on the one that we're going to end with, so there's a lot of advantages to working all by myself. But once I get it up to the point where other programmers can contribute, then I become a kind of technical team leader. Whoever is managing the people oversees all of us, and I am left to make any major changes in the core of the program while everybody else is working around the outside."

One of Bruce's colleagues in R&D was a high school senior named Seth Powsner (#108). He had just turned sixteen years old when he started working for MDSI, but he'd already been working at Comshare before that, and that's where he met Bruce Nourse. Seth remembers, "I was working on some Fortran / Xtran code to solve simultaneous polynomial equations by Newton's iteration, but Pioneer High School didn't have a high-speed printer suitable for such a long program." Bruce agreed to print it on Comshare's SDS 940 printer. Says Seth, "He offered a couple of tips, and encouraged me to keep on programming."

When Comshare was cutting its payroll in the summer of 1969, Seth was recommended to MDSI as an assembly language programmer, and Chuck and Bruce invited him for an interview. Seth worked full time for MDSI the rest of that summer and part time through his last year of high school. That was only the beginning of his MDSI connection. "I literally grew up at MDSI," he remembers.

"Seth was a whiz," says Chuck. Even though the young employee had only just gotten his driver's license, Chuck could see that Seth was as good as anyone on his team. He went to Ken and said, "I can't pay this kid like a high school kid. I have to pay him like my other R&D people."

As a part-time employee in the increasingly crowded Packard Road offices, Seth sometimes had to share a desk or borrow someone else's teletype machine (though he never had it as bad as his first "furniture" at Comshare, which was a large cardboard box). One of Seth's first assignments was to make improvements on the machine tool link for the Cincinnati 220 drill by programming it to do a "chatter routine" that would cut a circle. He later heard that sales of this link were even more popular now that it had this new capability. Says Seth, "So I established that I could do some programming."

One afternoon, Chuck gathered his R&D team to talk about a customer request that was proving to be a tough challenge. "We considered it for a while," Chuck remembers, "and we finally decided that all this effort would only help one customer, and we needed to be solving problems that helped all the customers. I told my guys, 'I think we have to turn them down.'"

Seth piped up and asked, "Can I try it?"

Chuck shrugged. "What have we got to lose?"

After staring at a blank chalkboard for a short while, Seth picked up the chalk and began writing quickly, covering the board with calculus equations. After a while, he asked Chuck if Don Willan could check

Seth Powsner in 1974, a little more grown up than the teenager he was when he first joined MDSI as a programmer in 1969. Photo: Bruce Nourse.

his work. Everyone knew that Don was the best mathematician at MDSI. Don found a mistake about three-quarters of the way through and erased everything after the mistake. Seth re-did the remainder, programmed it all in Fortran, and showed the results to Chuck. As Chuck remembers it, "We gave it to the customer in Texas, and two days later, they called to say, 'We have just made the finest pipe-straightening rollers in the history of this company.' I was so proud to tell them, 'That program was done by a high school student.'"

Without a car of his own, Seth came to work on his bicycle, or occasionally on his unicycle. He enjoyed the camaraderie of the MDSI basement crew (the self-described "cellar dwellers"): "It was all programmers and a minimum of interruption. The only problem was re-adjusting to sunlight when we came up for air." Seth remembers Bruce, Don Willan, Don Colley, and a fourth engineer playing bridge every day while eating their brown bag lunches. "The only difference between them was whether or not they refolded and reused the brown paper bag for a week."

Seth never forgot one of Bruce's strategies for minimizing interruptions. Sometimes the telephone on the teletype machine would ring, even though those lines weren't intended for phone chats. If Bruce's teletype rang, he'd pick up the receiver, say "Sorry, wrong number," and hang it up. Says Seth, "He didn't ask who was calling. The caller would apologize and hang up without realizing neither party had identified itself. It worked fine and saved time." (At some point, Seth, Bruce, and other programmers installed a second landline and a teletype machine at their homes, so they could program in the evening and save the prime-time use of the XDS 940 for MDSI customers.)

New hires to the R&D group were sometimes surprised to be introduced to their teenage coworker. This included David Hinckley (#139), who arrived toward the end of Seth's senior year at Huron High. Seth remembers, "Even though I had more XDS 940 programming experience than he did, Dave quickly realized I could use some practical advice in most other areas." Dave told Seth that he would never be taken seriously by adults if he didn't start wearing a necktie. Seth took the advice and started arriving to work in a tie, though still on a bicycle.

When Seth headed off to study electrical engineering at MIT, Dave gave him two other pieces of advice: "Don't let embarrassment stop you." And, "If in doubt, get a Penicillin shot." Seth returned to work at MDSI every summer through his college years. Secretary Betty Ruddy called him "the bad penny who keeps rolling back."

It was Bruce who told Seth that the first rule of programming is "Fix the first thing that's wrong first." After all, if a program's calculations are wrong, there isn't much to be learned from later routines processing bad numbers.

This turned out to be a great lesson when, after MIT, Seth went to Yale Medical School and began to diagnose patients. (Though Seth's focus had turned from programming to medicine, the staff in the med school soon learned they could turn to him for computer help.) Seth went on to become a board-certified psychiatrist practicing in Chicago until he returned to New Haven. (His wife, also from Michigan, had lived in New Haven and vastly preferred it to Chicago.) Seth went back to Yale, this time as junior faculty. He eventually became a professor of psychiatry and emergency medicine. He's known around the Yale campus today as the guy who always wears a bow tie.

■ ■ ■

By August 1969, COMPACT II was being used by several notable companies, including Naval Ordnance in Louisville, Kentucky, and Ex-Cell-O Corp. in Detroit, and it was getting some nationwide press in industry magazines. In addition, MDSI had added a regional salesman working out of an office in Caldwell, New Jersey.[13] Despite these early successes, the end of the first fiscal year (which was September 30, 1969, eight months from start-up) showed sales at just $29,000 and losses at $96,000. This might have made the investors gulp, except it was much better than the original projected loss of $157,000.

MDSI clearly still had a long way to go to be considered a viable company. But as Ken might put it, they'd had a successful takeoff. Now it was time to soar.

Part III
The Scale Up

———

10

Working the Business Plan

In a "President's Message" dated November 25, 1969, Ken compared MDSI to the other companies offering NC programming via computer timesharing: GE, Westinghouse Electric Corporation, and University Computing Corporation (UCC).

> *Both GE and Westinghouse have severe technical difficulties which make their systems much less than satisfactory for most applications. The UCC system is remote batch in APT and works quite satisfactorily but does have shortcomings that give COMPACT II a clear edge in ease of operation and in cost. As a result of these conditions, MDSI has a short-term opportunity to establish a much stronger foothold in the marketplace than originally anticipated. Plans are underway to expand the staff of MDSI as rapidly as possible to take advantage of these market conditions.*

Ken understood that technologies change quickly, that open doors can suddenly close. In response, he had hired salesmen to cover four regional territories—the Midwest; the East Coast; the West Coast; and Hawaii, the Virgin Islands, and the West Indies. As new salesmen came on-board, they racked up significant travel costs, while sales lagged well behind expenses. The investors saw the need for an urgent push, but everyone was feeling the pinch.

One way the team tried to save money was by limiting long-distance telephone charges. As Van remembers, "We had field people all over the country, and when they ran into a problem, they would call the home office person-to-person and ask for "Mr. Goodyear." He was never in the office, so the call would be disconnected, and there was no charge. But then someone in the office would call back on our less-expensive office line."

■ ■ ■

MDSI's business plan was built on several intended revenue streams. The first was connect-time: customers would be charged $10 an hour to connect to the timesharing computer service. The second was CPU time—the amount of time it took the XDS 940 to process the instructions of a part program: customers would pay $.30 per 0.01 minute of CPU time.

For any one job, these costs were easy for customers to manage. Early articles about the company gave example estimates. Said the *Metalworking News*, "The cost of programming an NC tape for a master fixture that takes over 630 boring and drilling operations would be $60 and would take 1.5 minutes of computer time."[14] *American Machinist* put it this way: "A point-to-point machining tape for a Milwaukee-Matic Eb, with three bolt circles (each different) totaling 54 holes programmed for drilling" resulted in a twenty-nine-foot tape that cost "$3.30 for 20 minutes of connect time and $12.90 for 0.43 minute of CPU time."[15] With more customers and more jobs—and the fact that MDSI would be paying Comshare only $.03 of every $.30 earned—these ongoing charges would begin to add up.

A third revenue stream seemed at first to have more immediate income potential: the sale of the machine tool links. Once customers saw the potential for COMPACT II in the shop, they would need to purchase customized programming to make the software work for their particular machine tools.

Chuck had originally figured that a machine tool link could be sold for $300 to $1,500, depending on the complexity of the code. But such upfront costs were proving to be a crippling deterrent to sales. "We discovered that no one wanted to spend that kind of money on the links," says Ken. "We were making plenty of sales calls, but we were closing very few sales."

By the end of 1969, Ken was beginning to question whether this whole endeavor could survive. Then one morning, while he was shaving, he had an epiphany.

When he arrived at the office, Ken called a meeting of all the employees, which by then numbered about a dozen. He explained the problem and concluded, "We're going at this the wrong way. What if we essentially give away the machine tool links and have a storage charge for each link of only $10 a month?" The customers wouldn't have to come up with $300 or $1,500 to seal the deal, but they would also never own the link; they would have to pay for it indefinitely. A factory with ten different machines would only have to come up with $100 in link fees to get started, which would enable MDSI to close the sale. For MDSI, that $100 would keep coming, month after month after month.

Chuck's initial reaction was shock: "How can we do that? We'll go out of business! This will sink the ship!"

But after some discussion, the team decided to give it a shot. To cushion the risk slightly, Ken also increased the CPU charge from $.30 per 0.01 minute to $.35.

Soon enough, customers were responding just as Ken had predicted. Bob Guise later called Ken's idea a "stroke of genius."

"Anybody who had any sense couldn't afford to say no," Ken remembers. "I mean, they couldn't hardly make a mistake at just $10 a month." MDSI touted their offer as basically "giving away" the machine tool link to encourage the customer to try COMPACT II. This low-cost entry decreased the time between the first sales call and closing the deal, and the total number of MDSI customers began to climb. In Ken's 1970 end-of-year report to the company's investors, he noted, "Effective March 1, 1970, MDSI revised its pricing structure by raising its usage prices but eliminating all initial charges including both software and training. As a result, monthly sales input per sales team has more than doubled."[16]

The first machine tool link that Bruce and his team wrote, which was for the Cincinnati Milacron 220 drill, was soon being leased to 200 different companies. Chuck remembers, "That generated two grand every month for just that one link." As each link was developed, its existence motivated other companies with the same machine to sign on. After a while, when an MDSI salesman walked into a factory, the customer would say, "Well, I've got a such-and-such machine," and the salesman could say, "No sweat. We've got that covered."

■   ■   ■

Carles "Cai" Raber, head of the MDSI team that wrote machine tool links, 1974. Photo: Bruce Nourse.

Creating the machine tool links became the responsibility of Carles "Cai" Raber (#109) and his small team of link writers. Cai was hired on July 1, 1969, after working several years as a computer programmer with a subsidiary company of GE. He had a master's in computer science from Dartmouth's Thayer School of Engineering, though when he graduated in 1962, that degree was mostly a study of electrical and mechanical engineering with just two computer courses. By the time he came to MDSI, he had more experience than most with the programming of NC machine tools. Cai was soon put in charge of the link-writing team.

Cai added Dave Hinckley to the team in April 1970, even though Cai considered it a "risky hire" because Dave didn't have a college degree. Cai remembers, "He was good at math, and he had a good, logical head on his shoulders, even though he didn't have the systems background that some of us came with. He really bootstrapped himself, and he did a really good job. He came a long way with his career in the company."

Cai quickly realized that customized machine tool links would be an ongoing request from customers. "You could assume that the link for a Kearney & Trecker mill with a GE controller would work the same for everybody, but customers are all different," Cai remembers. "They would try the link and then say, 'Well, I'd like to be able to make this other kind of part.' If we said, 'Well, we don't have something like that in our link library; you'll have to figure it out,' they'd say, 'No, that's your job.' Or they'd call us and say, 'I just bought a new version of this machine, and it has this new feature that your link doesn't cover.' Over time, we built up quite a library of links."

MDSI eventually ended up with more than 3,000 machine tool links. Says Chuck, "The result was that with each new customer, the probability of already having a link to run his machine kept going up and up and up." This huge inventory of ready-made links became another competitive advantage for MDSI.

■ ■ ■

A few other key hires in the first year helped push MDSI from a start-up effort to a professional and soon-to-be profitable corporation. Van was thrilled when MDSI hired Ron Pinnow (#110) in September of 1969 to take over COMPACT II training for customers. Ron had used the SPLIT language while working at Sundstrand Aviation, so he needed little training to understand COMPACT II. Plus, he already knew Van from Van's days at Lear Siegler (which had a contract with Sundstrand).

Another addition to the team was Frank Robert, whom Van remembers meeting while doing a demo with Chuck at a plant in Port Huron. Frank was one of the plant's part programmers. After the plant signed on and Frank started using COMPACT II, he applied to work at MDSI. He arrived November 3, 1969, and he eventually moved to California to take charge of MDSI's West Coast region. (One of Frank's former colleagues at the Port Huron plant, Frank Rose (#207), made the jump to MDSI in 1973.)

Chester "Chet" Fleszar was hired the same day as Frank Robert. They were employees #114 and #115—that is, the fourteenth and fifteenth additions to the team. Says Chet, "Betty Ruddy must have processed us alphabetically, because my employee number was lower than Frank's, a fact I never let him forget." Chet joined MDSI as an application engineer, but after just six months, he became the manager of the company's newly created customer support team—an ideal role for Chet's particular skills.

Raised on a farm near Saginaw Bay, Chet knew early on that he would not be a farmer. At seventeen, with World War II underway, he had convinced his dad to let him enlist in the navy. Chet became a flight engineer in combat planes flying over Europe. After three years,

Chester "Chet" Fleszar, in casual clothes for a company picnic, during his years as head of customer support for MDSI. Photo: Bruce Nourse.

during which he grew into a team leader, Chet attained the highest rank an enlisted man could get without more time in the military: Aviation Machinist Mate First Class. Chet credits that experience with developing the people skills that would help him excel in his career.

When he got home, he married his teenage sweetheart, Theresa, and went to night school to study engineering at the Detroit College of Applied Science. He got a job at Monaghan Bronze, a manufacturing company in Flat Rock, Michigan, that employed more than 700 people. When the company built a wind tunnel to test window walls for building construction, Chet remembers, "They needed someone to run it, and I had considerable aerospace experience." He stayed for thirteen years, working his way up to chief engineer and then director of engineering.

In 1969, Chet began researching numerical control. Monaghan Bronze didn't yet have any numerically controlled machines, but Chet knew it was coming. He talked to an NC instructor at Washtenaw Community College, Dallas Garrett, who told Chet to look up Chuck Hutchins. That summer, Chet and Chuck met at a numerical control symposium in Cincinnati. Chet remembers, "Chuck asked me if I would be willing to spend an hour or so with Ken Stephanz to explain my views on what manufacturing companies needed to be successful."

Ken and Chet had a long—four-plus hours—conversation on a Sunday afternoon. "We discussed a wide variety of topics," says Chet.

"We talked about the bigger picture of how American manufacturing was moving into these new technologies. I told Ken that I felt the future of the industry could not remain with those who merely turned the handles on the machines."

The result of that conversation, Chet remembers, was "Ken made me an offer I couldn't refuse."

■ ■ ■

Betty Ruddy liked to tease Chet that the only reason Ken hired him was because he was older than Ken. Indeed, at age forty-four, Chet was the oldest member of the growing MDSI staff. That maturity came through in his management skills and his understanding of what NC meant to the machinists who would be trained to use it.

He decided to focus on customer support because he "recognized that the need was there," Chet remembers now at age ninety-six. "I was out in the field among customers and saw what their real problems were in the use of our software." Those real problems included not only technological challenges but psychological ones too.

Chet was particularly concerned about the part programmers who may not have been involved in the decision to add an NC machine to a shop but were now expected to make it run. "It was somewhat difficult for them to grasp the concept that this was not going to take their jobs away. This was merely going to give them the opportunity to do what I considered to be very important, and that was time to think. I borrowed that from Henry Ford, who said, 'Thinking is the hardest work there is, which is probably the reason so few engage in it.' Most of them were very skilled machinists, but they just didn't have that grasp of what these new technologies could make possible."

Chet remembers one company in Green Bay, Wisconsin, that signed up with MDSI when Chet was still an application engineer. "It was a small company, just one part programmer. The machine was rather complicated, but everything was going fine. Then about three months later, the plant manager called and said they weren't getting the productivity they expected." The manager thought there must be something wrong with

COMPACT II, and he wanted MDSI to fix it. Says Chet, "I visited the plant, and I asked permission to go talk to the programmer. I chatted with him for a while, and I learned that he used to work six or seven days a week, but now he was working just forty hours a week." The machinist viewed the arrival of the NC machine as causing a cut in his pay.

Chet went back to the plant manager and asked, "Did you consider giving him some increase in salary when he took on the added responsibilities of programming?" When the manager said he hadn't, Chet said, "Well, I think I just showed you what your problem is." On a subsequent visit two months later, says Chet, "The programmer was all smiles." And productivity was up.

That kind of service wasn't necessarily the intent of MDSI's customer support, but Chet understood that change is complicated. "I had more experience in industry and management than most of the early employees at MDSI, so I could see when our customers were going through the pain of adopting the new technology." He regularly talked to machinists who used to feel completely confident in their jobs but were now frustrated by this new role. And it wasn't just the part programmers who felt that pain. "Management in many of these companies did not recognize what this new technology was going to do. All they knew was they had to do it because of competition."

Of course, customer support also handled practical calls about using COMPACT II to create punched paper tapes. Comshare's online system made it possible for MDSI's employees to review a customer's part program in real time, resulting in instantaneous troubleshooting. This service was explained in an issue of Comshare's newsletter: "The COMPACT II programmer can obtain instant help at his terminal if he runs into a problem during processing. He simply calls MDSI on the telephone and asks that an NC engineer log in on a terminal located at the MDSI office. This done, the MDSI engineer can link his terminal to the user's terminal for an online conversation. Whatever is printed on one terminal appears on the other, making it possible for the specialist to watch the user's program in action."[17] In other words, MDSI offered its customers remote-access IT support before most people knew what that was.

As the start-up passed its first anniversary, sales were increasing, but so were expenses. Ken's prediction that the company would not become profitable until sometime in Year 2 was, so far, holding true. And Ken knew they needed to keep growing to have a chance of succeeding.

The company quickly burned through the first installment of investment dollars, but when it came time for Comshare to put in more money, Bob Guise pulled out of the deal, citing financial difficulties.[ix] It was a tough time for all tech firms following the tech stock crash of April 1970. The S&P 500 and the Dow fell 19 percent over five weeks. The average computer stock fell 80 percent from its peak value in late 1968 to the May 1970 low.[18]

Morgenthaler's right-hand man, Bob Pavey, remembers that his boss was vacationing in Italy when the tech stock bubble burst. "I sent Dave a letter, saying 'MDSI needs money.'" Meeting that need would require new investors.

Bob Pavey was on the MDSI board by this point, and he remembers, "My value was analytical, to really get in there and look" at the numbers, the business plan, the strategies. "Because this was a major investment for us, I got close to the company. I went to planning sessions. I was the pushy MBA punk." What Pavey found was that MDSI still had great potential, but it needed more funds to grow into a profitable enterprise.

Ken reached out to George Simon, owner of US Equipment Co. in Detroit (and the creator of the record-breaking hydroplane power boat *Miss US 1*). Simon was, at that moment, in the Caribbean, but he took Ken's call, listened to the pitch, and said that he would check things out as soon as he got back to Michigan. He soon became an MDSI investor and joined the board.

Dave Morgenthaler then turned to California-based investors Bill Hambrecht (of Hambrecht & Quist, one of the first major investors in Apple and Adobe) and Philip Arthur Fisher (the author of the 1958 investment guidebook *Common Stocks and Uncommon Profits*, which

---

ix In 1970, with Comshare in financial trouble, Bob Guise and Comshare parted ways. Comshare co-founder Rick Crandall then became Comshare CEO.

is still in print to this day). He also reached out to a YPO colleague, Seth Atwood, who was a Harvard MBA and auto body parts supplier in Rockford, Illinois.[x]

All three men did their due diligence and agreed with Dave that MDSI looked like a good bet. The total venture capital investment in MDSI was $1.5 million (about $10 million in 2020 dollars), and it proved to be just enough to get the company into the black. When Morgenthaler and his fellow investors were eventually well rewarded, "A little venture capital firm in Cleveland," says Pavey, "got to be well known because we backed this amazing company called MDSI."

x   Atwood was also a collector of clocks and timepieces who founded the Time Museum in Rockford, Illinois, one of the leading horological museums in the world from 1971 until its closing in 1999.

11

A Professional Company

In the spring of 1970, MDSI partnered with Cincinnati Milacron for a four-city tour of NC users. Chet Fleszar remembers giving two live demos each day, showing off both COMPACT II and the NC machines manufactured by Cincinnati Milacron. After a demo at a New York–based manufacturer, the sales team was about to head to dinner when someone from a New York electric power company asked the Cincinnati sales manager for help.

Says Chet, "It seems they were having a serious problem at one of their main control centers and were in dire need of a critical part they called a 'fuse.' They had the right material on hand, but not the equipment to make it." The part was fairly difficult to produce, but the MDSI team quickly wrote a program, returned to the demo area to enter the information, and produced the tape. After the part was machined, the power company men inspected it thoroughly, thanked everyone profusely, and left. Chet remembers, "The next day we were informed that our action probably saved some parts of New York State from a blackout."

Missing dinner—or a whole weekend—to make a sale wasn't at all unusual in the early years. Chet remembers another 1970 demo he did in Boston. When the demo was done, the plant manager invited Chet to an early dinner out on the Wharf. They had just been seated when Chet was told he had a phone call. It was one of the new MDSI secretaries, Judy Foster Leverett (#144), on the line. She told Chet, "Ken wants you in Colorado tonight for a demo tomorrow morning. Your flight is all arranged. You can pick up the tickets at the airline counter."

Chet looked at his watch. He had barely enough time to get to the airport and make the flight. His host rushed him there, promising him a rain check on dinner.

Chet arrived in Colorado late that night, met up with Ken the next morning, and with little time to prep, they did a demo on a Hughes Dual Center (one of the most complex early NC machines). "I thought the demo went well," Chet remembers, "and after we finished, Ken carried the terminal all the way out to the parking lot and honored my open rain check by taking me out to dinner."

The hectic schedule could certainly get tiring. One evening, after three grueling days of manning the MDSI booth at the International Manufacturing Technology Show in Chicago, Chet and a new application engineer (AE) named Bob Drewry (#131) drove to Midway to meet up with Ken, who was going to fly them back to Ann Arbor. Ken wasn't in the airport terminal when they arrived, so Chet and Bob sat down to wait for him. When Ken walked in a short time later, he found both men sitting up in the airport chairs, still in their suitcoats and ties, sound asleep.

Such were the demands of a fast-growing company.

Ken's high expectations pushed MDSI out of the start-up phase and into a period of rapid growth driven by professional standards. He remembers, "We tried to be a professional company in terms of behavior, attitude, the way we conducted ourselves, and so forth. That seemed to permeate all through the company."

MDSI application engineer Robert Brown (#171) (on left) reviewing a customer's Compact II part program (with a DeVlieg JigMil in the background), 1977. Photo: Urbanes Van Bemden.

Customers felt it too. If they reported a software bug, Chuck's R&D team or Cai's link-writing team fixed it the same day. After all, factories often ran three shifts; even a few hours of a machine being down could cost a company greatly. MDSI was all about making things better for their customers; they didn't want to be in the business of making things worse.

Not that the software engineers were always the most professional. "They could be wacko," Ken laughs today. "I'd walk into their area and find guys lying on the floor, with their feet up on the chair or desk and the keyboard on their laps, writing code." Ken may have set the tone with his high-back chair and business attire, but he had an engineering background. "I understood the engineer's mind. I may not have understood the technology at that point, but I knew how they thought."

■　■　■

One of those engineers was Bob Drewry, who arrived in March 1970. He turned out to be a key hire, not just because he was a good application engineer, a capable programmer, and, as Ken puts it, "a real solid citizen," but also because he had a head for accounting. Those skills enabled Drewry to make a significant computing contribution to MDSI that had nothing to do with machine tools.

For more than a year, Ken had maintained meticulous financial records by calculating, with a slide rule, the company's ever-changing figures, hand-writing them on ledger sheets. The sheets grew longer and more complex as the business grew increasingly complicated. Eventually, Chuck said, "Ken, I just can't bear to see you doing that. Let me ask Bob Drewry if he could write you a piece of software to help. He's surprisingly good at writing Fortran code."

Chuck then asked Bob if he could write a program that looked like a big sheet of column paper to help Ken project MDSI's income and expenses, given a certain set of assumptions. Bob was willing, but at that time, no one knew the term "spreadsheet." This was breaking new ground. Bob sat down and began to envision how he could program a two-dimensional matrix (array) in Fortran. The first subscript would be the row, and the second subscript would be the column. Each position in the array would hold a value, either assumed or computed. The column upper limit would be sixty—one column for each month for five years of projection. The row limit would depend on how many assumptions the company wanted to make and how many different variables needed to be projected. After a little planning, Drewry told Chuck, "Yes, I think I can."

Primarily, the categories to be projected were income, expenses, cash flow, and profit. The basic income assumptions (number of customers signed per salesman per month, revenue generated per customer per month) and expenses assumptions (expenses per sales team per month, expenses for other positions in the company) were similar to those Ken had plotted when he developed his first detailed cash flow projections. After more than a year in operation, MDSI could now make other, more accurate assumptions about costs, including how many customers could be serviced by one application engineer before another AE was needed in that territory, how long it took for a new salesman to become productive, and what support positions were needed as the number of customers and employees grew. The company could even make fairly accurate assumptions about the cost of office furniture for new employees, increasing telephone charges, and so on. These were the kinds of categories Bob Drewry put into his program.

"It was my first exposure to business accounting," he later remembered, "and I learned a great deal from this project." After about two weeks of focused work, Bob and Chuck demoed the program for Ken, who was sitting at the ASR33 teletype terminal. Bob had called the program "Financial Projections," but he named the executable file KEN for the purpose of the demo.

Bob thought he was being clever when he told his boss to start the program by typing, "EXECUTE KEN."

But Ken knew that the command dispatcher only needed the first three letters, so he typed EXECUTIVE KEN. That command worked just as well, of course. Ken had the last laugh.

But he wasn't laughing when he began to see the power of the program. After he specified a value for each assumption, the computer processed the calculations, and a report was printed. Ken saw that he could test a lot of "what if" scenarios in a short amount of time.

Drewry pointed out, "All the logic and computations are hard coded, so if you need to change any of the calculation methods, we'll have to edit the program." And a detailed printout required a lot of cutting and taping to produce a presentable row–column spreadsheet. Plus, it only made sense to do the computer processing in non-prime hours (after 5:00 p.m. and before 7:00 a.m.) when Comshare didn't bill MDSI for computer use. But even with these limitations, Drewry's program had significant time-saving potential.

After Ken gained some familiarity with the program, he asked Bob Pavey—the MDSI board member from Morgenthaler Associates—to audit the program for accuracy and methods. Pavey came into the office and worked with Bob Drewry for two days, running the program with different assumptions and analyzing the results. Pavey found a couple of minor errors, which Drewry immediately fixed. After that, Drewry remembers, "Our Financial Projections program had the blessing and confidence of our investors." Ken and MDSI's financial manager, Jon Ehrmann (#227, hired in 1973), used the Financial Projections program—with occasional updates and expansions—until 1980.

Chuck likes to say that Bob Drewry created the first-ever computer spreadsheet. As it turns out, a University of California–Berkeley professor—economist and mechanical engineer Richard Mattesich— had already developed an accounting spreadsheet program using the Fortran language. But no one at MDSI was aware of Mattesich's mid-1960s advancements when Bob Drewry wrote his program. For all intents and purposes, Drewry independently developed the first digital accounting spreadsheet.

The first successful commercial spreadsheet program (VisiCalc) did not come out until the fall of 1978, seven-plus years after MDSI began using the Financial Projections program. VisiCalc ran on an Apple II computer, which had just 48 KB of memory, the same as the XDS 940. VisiCalc was an improvement over Drewry's program in that users could change the formulas as they worked, and of course, the Apple II had a graphic screen so users could see their grid in real time as they built it. Lotus 123 (which came out in 1983) would become the next big spreadsheet program, and then Microsoft's Excel (introduced in 1987) became predominant. Drewry jokes, "With Excel, you could place the equivalent of our entire program in a single cell."

Looking back on his work, he reflects, "I have often wished that we had recognized the commercial potential for this program and started another development project. We could have been the original Microsoft Excel. We were really ahead of the curve. But the important thing then was to make MDSI a success."

■ ■ ■

By mid-1970, with eighteen employees on site, the Packard Road offices of MDSI were bursting at the seams. Ken found office space for rent at 320 North Main Street in downtown Ann Arbor, and that became MDSI's home for the next six years. "The move was most welcome," says Chet. "We needed the space and the organizational structure." And the R&D team needed to get out of the basement.

MDSI's staff was also growing throughout the country. Every quarter, the sales team was adding another regional salesman who lived in the area where he worked. "We hired direct employees only," says Ken. "We didn't work with representatives." MDSI gave each new salesman a car, assigned him a monthly sales quota, and sent him out to every factory within a certain radius. Very often, Ken was renting a single-engine plane to fly to each region and provide on-the-ground support.

Back in Ann Arbor, Betty Ruddy held down the fort on Ken's behalf. When one of the AEs returned from a trip and asked Betty what to do with the S&H Green Stamps (kind of like frequent-flyer points) that they'd received after buying gas or food, she put out her hand and claimed, "They belong to the company!" Betty saved the stamps, using them to reward good work. The staff started calling them "Stephanz & Hutchins Green Stamps."

Judy Foster Leverett felt lucky to be working as a secretary under Betty's tutelage. Judy was just twenty-six when she arrived at MDSI in June 1970. She was raised on a Michigan farm where her family grew five thousand tomato plants every summer. "I had waited on customers at our farm stand," says Judy, "so I'd grown up a lot already." She took some college courses and then applied for a job through a local employment agency, which eventually placed her at MDSI.

Judy remembers working in a group of three women—all secretaries—in a common area out in the middle of the workspace. The atmosphere was "friendly but always professional." Judy was expected to wear a skirt or dress (or occasionally slacks, but never jeans). The secretaries supported each other and did whatever work needed doing. Judy felt that everyone treated her as an equal. "People always said 'thank you.' I felt appreciated."

In her early years at MDSI, Judy served as Chuck's secretary, helping to arrange his travel, taking notes at meetings, gathering data for reports. She remembers Chuck as "Mr. Enthusiastic": "He was a nice guy, very knowledgeable, with *lots* of energy." Later, she worked as secretary to CFO Jon Ehrmann.

Judy liked that Bruce Nourse always had a joke to share and that Cai Raber was always even-keeled. "I don't think I heard any of those guys ever raise their voices." She often joined a group of coworkers for lunch at a downtown restaurant near the Main Street office, and she looked forward to the annual company Christmas party at the Campus Inn Hotel (around the corner from the office). After MDSI hired a link engineer (and guitarist) named Glen Pingston (#184), he would bring his band to play at the party so everyone could dance.

"It was such an adventure," Judy says of her time at MDSI, "to be there at the start of a new idea and to work with people who were so creative. If I had a chance today to work with people like that again, I'd do it."

But a friendly, professional, and tech-savvy company can only last so long without profitability. That necessity still eluded MDSI at the end of their second fiscal year, September 1970. What they needed were more customers.

12

The Salesman

When Ken Stephanz was considering hiring Mike Long as vice president of marketing, he learned that Roy Winn (#105) had worked with Mike at Bendix. Ken asked Roy, "What kind of guy is this Mike Long?"

Roy said, "He's an iron-assed sales manager."

Ken slowly nodded his head. "That's just what we need."

Mike Long (#148) arrived at MDSI in September 1970. He was thirty-five years old, and he had already started and sold a scientific instruments company, and had learned about NC machine tools at Bendix. He had a sharp mind, a serious demeanor, and, most importantly, a knack for sales.

■ ■ ■

Born and raised in Chicago, the eldest of three boys, Mike was the son of a man who had advanced from shipping clerk to general superintendent at Central Steel and Wire Company, a multi-state steel distributor based in Chicago. Perhaps it was conversation around the dinner table or on the golf course with his dad that piqued his interest, but young Mike found he wanted to understand how products are packaged and promoted and sold. When he left Chicago for the University of Colorado–Boulder in 1952, he decided to major in marketing.

Halfway through his college education, Mike was drafted into the U.S. Army. He was sent to study electronics in Texas and then to maintain early warning radar systems in Germany. He was also selected for the Noncommissioned Officers Academy in Munich, from which he was an honor graduate. After two years of military service, Mike returned to Chicago in the fall of 1956 and took a job as an electronics technician for Packard Instrument Company, maker of scientific instruments for medical research. When he attended scientific conferences to demonstrate the company's equipment, his instinct for sales and marketing resurfaced. He intended to go back to school, but his father died in the spring of 1957, and then Packard Instrument offered Mike a job as a salesman, covering the Midwest territory.

"I was very good at it," Mike recalls, "good enough that the president of the company appointed me as a regional manager and sent me to New York to open an East Coast operation, which I did. That turned out to be quite successful also."

Reflecting on his sales acumen, Mike says, "I always thought of myself as a retailer, not a wholesaler. I can be very convincing one-on-one. If I have a product that I believe in, and circumstances that provide a unique selling proposition, and if I can convince you that what I'm telling you is true, the only logical thing for you to do is buy my product. I sold a solution to a problem, simple as that."

As he sought to solve problems for Packard Instrument customers, Mike became aware of the need for an instrument that Packard didn't make and wasn't interested in making. He remembers, "I got into discussions with the chief engineer for the company, and he and I decided we'd go off and make it ourselves." They started Vanguard Instrument Company, which specialized in equipment used for paper and liquid chromatography. "When we started, we didn't have any money to speak of," Mike remembers. "I basically lived in my car as I drove around the United States, trying to sell this instrument. It took me about four months to make the first sale, but then we started selling a lot of them."

Four years later, Vanguard was acquired by Technical Measurements Corporation of Connecticut. By then, it was the early 1960s, and Mike—married with two daughters—suddenly had enough funds to step back and consider his next move. He finished up his undergraduate degree at Colorado and then earned an MBA at Northwestern focused on finance and marketing.

After that, he moved to Michigan for a position as sales manager with the Industrial Controls Division of Bendix. "That was my introduction to numerical control," says Mike. "The Industrial Controls Division was a primary control supplier to machine tool builders who were servicing the aerospace industry." Mike learned about NC programming through his work with a team at Bendix that ensured the controls they manufactured were compatible with the APT programming language.

As sales manager, Mike called on customers at Boeing, Convair, Douglas Aircraft, and McDonnell Aircraft. "As a matter of fact," Mike remembers, "I had the pleasure and privilege of walking through the first mockup of the 747 at Boeing's plant in Seattle." But Mike's particular role at Bendix was undervalued. "Bendix was never noted for its marketing prowess," he says. When the company's

Mike Long, MDSI's vice president of marketing, 1973. Photo courtesy of Mike Long.

leaders had to make some cuts during the 1970 recession, they decided they didn't really need the sales team.

Late that summer, Mike's search for new employment led him to MDSI, to Chuck and Ken, and to Dave Morgenthaler, who told him they needed a sales executive who could get the company profitable, and fast. "I looked at the company as best I could," says Mike. "It was very young and fairly opaque. I tried to assess what it was they were attempting to do and what wasn't proving successful."

When he joined the team, "things were pretty grim," Mike remembers. "We just weren't getting enough new customers." But very soon, Mike was living up to his nickname as an "iron-assed sales manager."

■　■　■

As he got to know the existing sales team at MDSI, Mike discovered they were mostly former machine tool salesmen and that such a background wasn't appropriate to sell COMPACT II. In general, Mike found that machine tool salesmen "did not have an aptitude for the kind of selling I had in mind. To be brutally frank about it, machine tools are not sold, they're purchased. People buy a machine tool when they have a need to buy a machine tool. All the machine tool salesman does is try to get the customer to buy his machine tool."

Selling the services of MDSI was an entirely different kind of sales challenge. Experienced machinists in shops that were new to NC machines were often frustrated by the programming difficulties they encountered. Without access to the large computer systems required to run a program like APT, they would often default to time-consuming manual programming. MDSI's salesmen had to convince them that COMPACT II's user-friendly programming language and timesharing computer system would not only save them from frustration, but would also save their bosses time and money.

Mike made two strategic changes to MDSI's sales approach. First, he gradually recruited a new team of salesmen who were "very bright, very aggressive people," none of whom knew about machine tools. One had sold Yellow Pages advertising; another had sold appliances at Sears. "They were intelligent," Mike remembers, "and they bought into my concept of what the job was like." Over time, he replaced essentially the entire sales organization. A two-week training program taught these men about machine tools and numerical control and why COMPACT II—with its use of words that were familiar to machinists— was a better solution to the programming problem than anything else in the marketplace at that time. Next, the salesmen were assigned a region and given quotas for number of new accounts, number of machines under contract, and revenue produced by those accounts. Mike held his salesmen to high expectations.

Concurrently, he created two-man teams—a sales engineer and an application engineer—to work together with potential new customers. Says Mike, "An aggressive salesman could do the cold calling and qualify the prospects and get demonstrations set up, and then he would bring in the application engineer to prove the concept." Although Ken had initiated the idea of sending AEs into the field to present to prospects, that usually didn't happen until a "sales engineer" (SE) called the home office and asked for an AE to fly out for a demo. It was Mike's idea to pair up a salesman and an AE to work a region together. "We very quickly settled on the concept of the sales team, so that, out in the field, all the time, the salesman was working with his application engineer."

MDSI's AEs were experienced machinists who knew COMPACT II inside and out, and could program almost any part with COMPACT II. At the demo, the AE could ask the machinist—who might be doing manual programming from blueprints—to talk about how long it had taken him to program a certain part that the shop was manufacturing. Then the AE could get on the portable teletype machine, write the program using the blueprint, process it, debug it, and produce the tape to make the part, right there and then.

Mike remembers, "The resulting part was identical to what they had done previously, but usually in about one-fifth or even one-tenth of the time. Now, if I've done that for you, how could you *not* give me an order? Having shown you the truth of what I'm telling you, you owe me the business; in your own best interest, you owe me the business." Mike adds, "As the results showed, most of them agreed."

Chet Fleszar still thinks the two-man sales team was a breakthrough for MDSI. "This was not only the best way to market our product, but it presented an excellent image of MDSI in the marketplace. The teamwork that was the hallmark of MDSI was a main factor in our success."

Van agrees. He liked that the sales engineer and application engineer always showed up looking like a couple of IBM executives in suits, ties, and white shirts. But it was their knowledge and service that got the most attention. "Our customer support was second to none," says Van, who notes that the two-man team was later copied by MDSI competitors.

■ ■ ■

For every demo, the AE brought along an ASR33 teletype terminal in a fiberglass case that was fondly called the "Blue Goose." MDSI had purchased these cases to protect the ASR33 during transport, but if the teletype terminal had to be checked as luggage for a plane flight, it often failed to work properly upon arrival. Van remembers sitting on the tarmac with SE Dick Weeden (#194) before a flight to Denver and watching from the window as baggage handlers loaded the plane. "Lo and behold, there was a baggage handler loading the Blue Goose, and we watched as he simply tossed it a few feet away onto the conveyor belt. Sure enough, it didn't work when we got to Denver." Usually, the problem was that the H-plate had come loose. Any local telephone company could fix or replace a broken ASR33 terminal, but Van and the other AEs much preferred to save time and fix it themselves.

Another tool that AEs started bringing along to demonstrations was a graphics plotter (a small machine with a roll of paper and a stylus). Though 3-D computer-aided design software was not yet a reality, COMPACT II could communicate with a graphics plotter that would draw a picture of a part based on the part program. Says Mike, "The plotter would actually draw an outline of the part as the program was being processed, so you ended up with a picture that you could compare to the blueprint and say, 'Yeah, that's accurate; that did the job.'"

Mike recognized that customers wanted these graphics plotters in order to prove out the program before they started to cut metal. "I negotiated an arrangement with Hewlett-Packard whereby they gave us 'permanent loans' of Hewlett-Packard plotters that our sales teams carried out in the field. Hewlett-Packard later told me that we became the largest influencer in the country of the purchase of their plotters." (He remembers how he tried to convince another company, Tektronix, to "loan" plotters to MDSI as a way of promoting the Tektronix plotter, but that company couldn't be convinced to make the deal. Says Mike, "Over time, that decision probably cost them a couple million dollars.")

■　■　■

To generate leads for his new team of aggressive salesmen, Mike implemented an advertising program. "Initially, I wrote the ads and got them placed," he remembers. The ads were designed with reply cards, which led to qualified leads.

Mike also sought free publicity by pitching articles to trade magazines. "There were two media sources that were opinion leaders, if you will, in the industry: *American Machinist* and *Metalworking News*. I would periodically go to New York and meet with the editors and try to get some publicity, which did a lot for our credibility in the early stages. Nobody knew MDSI. Those articles helped."

The sales teams generated other leads by getting to know the local machine tool salesmen in their regions. Says Mike, "That was a kind of symbiotic relationship, because the machine tool salesmen knew that MDSI would make it easier for their customers to use the NC machine tools they were trying to sell."

Mike was based in Ann Arbor, but he spent a lot of time flying around the country to sit through demonstrations, to assess and coach the sales teams toward greater success. "I was on the road usually four or five days a week, every week." After about two years, he received a plaque in the mail from United Airlines, commemorating his half-a-million miles in flights. Laughs Mike, "I hung that up on the wall."

His demanding schedule was expertly managed by secretary Carol Guttman (#224), whom Mike hired in July 1973. "She was a New Yorker, and she was tough as nails," Mike remembers. Carol was a graduate of Vassar College, a mother of seven children, and an active community volunteer. "She was a wonderful person. She was my strong right hand, and she helped me very much over the years."

With more publicity, increasing and more qualified leads, strong sales teams, and good leadership—combined with the right pricing model—sales began to rise at an ever-increasing pace. Chuck remembers, "Mike built an incredible sales organization. We hired Mike, and that began the company's real acceleration."

One year after Mike's arrival, an article in *Metalworking News* noted, "In the 31 months since COMPACT II was developed, MDSI has 250 customers running 870 NC machine tools."[19] Six months after that, Mike told the *Metalworking News*, "At the latest count we have 325 customers and 1,250 machine tools utilizing our service. That means 5 percent of all NC tools in place in the U.S. are on our COMPACT II. That makes us the most universally used language…rivaling even APT, since there are so many versions of that."[20]

Mike remembers that, when he first arrived at MDSI, Betty Ruddy kept a bell on her desk that she rang every time a salesman brought in a new order. "She'd walk through the office, ringing this bell. Unfortunately, the bell wasn't ringing very often at the time I joined. But after about a year, she was ringing that bell so often, it was rattling the office all the time, and she basically had to stop doing it."

13

Customer Support

By February 1973, MDSI had 500 customers running about 2,000 machine tools. According to industry leaders, the nation's entire complement of NC machine tools at that time was no more than 22,000. Thus, just four years after its founding, MDSI was serving almost 10 percent of the entire American market.[21]

As customer expectations for COMPACT II grew increasingly sophisticated, the R&D group would hear feedback from the AEs whenever they gathered in Ann Arbor for national sales meetings or training. Chuck and Bruce always hosted a session at these events to invite the AEs to share their wish lists for COMPACT II. Says Bruce, "People in the plants gave the AEs ideas of things that would improve the program, that would make it more versatile."

The problem was that MDSI's R&D team kept bumping up against the memory limitations of the XDS 940. Van remembers that Chuck or Bruce valued the feedback but always had the same comeback: "We only have six instructions left. We don't have room to add that."

Everyone was frustrated. "Finally," Van says, "they broke the code and came out with syspages. It was something magical." A syspage was a set of instructions for a task that was always done the same way, like all the moves and cuts necessary to turn a shaft.

Bruce explains, "There was only room for eight pages of code with the XDS 940, and each task required a big block of code. But one block of code doesn't need to be running while the computer is using a different big block of code. We called those big blocks "syspages," and we could swap syspages in and out of the eight pages. That freed up memory so we could program more functions."

Van summarizes, "That solved a lot of our problems. To be quite honest, I don't know how the system people got so many capabilities out of such a small amount of core memory."

Some of these syspages became new products that MDSI could

promote. For example, "FasTurn" was introduced at the 1971 Numerical Control Society meeting. Mike Long told *Metalworking News* that FasTurn could, in some instances, reduce "forty or more program statements to just three: definition of the material boundary, the part boundary, and one statement for the actual cutting. These three statements can generate up to 500 or more individual tool motions."[22]

Seth Powsner remembers, "Don Colley programmed the logic behind FasTurn, and Don Willan wrote the code for pocket milling which we called OptiMil. Both of these were impressive to me, even as a student taking computing courses at MIT. Don Colley would joke that his next syspage was going to be CUT-THE-PART. I kept wondering whether I'd return some summer to find that he had cracked the problem and had literally added a syspage to go straight from geometry to a finished machine tape."

■　■　■

Another request from customers was that COMPACT II accept instructions in metric measurements. Chuck remembers well the day Mike came to him and said, "The world is going metric. We now have drawings in centimeters, even if most of the machine tools still operate with inches. But COMPACT II only works with inches—inches in and inches out. We need to be able to take metric in and put inch out, or take inch in and put metric out."

Chuck let out a long sigh. He knew it was true that customers were going to need this capability. But the required code would have to be added to Page SIX. And Page SIX already had 2,048 instructions in it. There was no extra space.

Bruce Nourse wrote the code that would make the switch between inches and metric measurements. It wasn't too complicated: just twenty-one lines of code. But where to put them? Chuck put his seven-man team on the task. They needed to review all 2,048 instructions on Page SIX and find twenty-one places where the code could be simplified, combined, or rewritten to remove a line of code. "It took seven guys seven working days—forty-nine man-days—to find those twenty-one locations," says Chuck. "It was the single biggest software problem we had to solve with COMPACT II. But we did it." COMPACT II's metric capabilities were announced in March 1972.[23]

■   ■   ■

The first thing every new customer needed was training in how to use COMPACT II. The training was free, but the customer had to send their part programmer to Ann Arbor for a two-week course. These courses occurred almost continuously, so a customer would not have to wait long to start using the new system. In time, MDSI was holding these trainings in a local hotel ballroom to accommodate the size of each class.

After the training, the part programmers returned to their manufacturing companies ready to use COMPACT II and whatever machine tool links had been customized for their particular machines. But questions and problems were inevitable, and customers called MDSI every day for help with programming challenges. That's where Chet Fleszar's "customer service engineers"—the CSE group—took over.

At first, it was just Chet and Don Young (#127) answering the phones, with Judy Foster Leverett as the group's secretary. In mid-1970, Chet brought on the first female engineer at MDSI, Donna Bohling (#146). "That did raise a few eyebrows," Chet remembers. "But it worked out successfully." Other key CSEs, like Earl Chadwick (#222), Terry Peterson (#288), and Frank Stancato (#305) came on a couple of years later. (Chet found some of his employees through a course he taught at Washtenaw Community College. Ken had encouraged Chet and a couple of other senior managers at MDSI to give back to the community by teaching.)

Eventually, CSE became the ideal training group for future application engineers. They would spend about a year working the phones and providing customer service, and when they knew just about everything that customers were trying to do, they were promoted to AE and sent out to work hands-on with customers in the field. Chet estimates that, over the history of MDSI, at least a dozen AEs were trained in the CSE group and then transferred to field positions. These included another female CSE, Ann Cline (#649), who became an AE in 1978.

When MDSI added an 800 number so that customers could call in for free, it was a bold move for the times, since MDSI would be covering the cost of those long-distance calls. The ability to offer customers "toll-free" calling was a relatively new service in the early 1970s. As anticipated, customer traffic did increase initially, creating more work for the CSE group and more costs for the company. "But the positive feedback about our professional way of doing business was well worth it," says Chet. The CSE group set a goal of "no calls being escalated"—they wanted to do their job and solve customer problems themselves. Most days, they met their goal. "It was a result we all relished," Chet remembers, "because we had done our job."

To keep customer support at a high level, CSE developed a call classification scheme with code numbers for each type of customer issue. This sped up the documentation process and resulted in a database of coded problems that included how much time it took to solve each problem. Says Chet, "I used that in training our future application engineers, because now I could see what most of the customers were having a problem with across a range of codes and how much time it should take an experienced AE to fix it. This allowed us to monitor the progress of our itinerant staff of field AEs. Plus, we could transmit the information to our software engineers so they could improve the product."

In reflecting on this approach, Chet says, "I thought it was my responsibility, as a manager, to create something like that. It was these kinds of services—like building a database out of customer problems—that drove the success story of MDSI."

■   ■   ■

By the end of 1970, MDSI had run through the bulk of its venture capital. Ken had predicted a break-even budget by the twenty-second month, but, in fact, it took twenty-four. "Finally, we broke even at the two-year mark," Ken remembers, "but only after making our investor partners very, very anxious!"

Bob Drewry was on a work trip when the "positive cash flow

celebration" was held at MDSI headquarters. "But some good souls saved a bottle of champagne, and when I got back, they celebrated with me."

Increasing success brought increasing challenges, of course, especially for a company providing services within the quickly evolving computer industry. One of MDSI's most critical challenges was securing enough computing power to meet the needs of their expanding customer base.

Comshare was MDSI's primary provider for the first couple of years, eventually purchasing ten XDS 940s to meet MDSI's demand. But as Comshare tried to grow into other markets—primarily providing timesharing and mainframe database technology for business management—it was unwilling to promise dedicated computers to MDSI, even though MDSI was Comshare's biggest customer (and originally a part of Comshare). According to Ken, "We got the short end of the stick with Comshare, and as time went on, it grew shorter and shorter. Any deal I ever negotiated had to be fair for both parties, but they weren't dealing fairly with us."

In frustration, Ken turned to Tymshare in May 1971 for additional computing capability, as well as a third company, Computer Complex, Inc., of Houston, Texas. By September 1971, MDSI was using nine Comshare 940s, two Computer Complex 940s, and sixteen Tymshare 940s.[24]

Unlike Comshare, Tymshare was willing to offer computers dedicated solely to MDSI customers. For several years, Tymshare provided a new, dedicated computer to MDSI every six months. Plus, Tymshare had a larger network of XDS 940s in locations where Comshare's network didn't go. "We ended up being Tymshare's biggest customer," says Ken. He estimates that Comshare eventually lost many millions of dollars in revenue to Tymshare.

■   ■   ■

One market that put MDSI over the top financially was manufacturers of oil well tools. Ken wanted to put MDSI's best resources to this lucrative customer base, so he asked Van to move with his family to Texas to serve as the field AE for that region.

Van was teamed up with a sales engineer based in Fort Worth named Jim Spencer (#140). "He was kidded a lot," Van remembers, "because he was a former perfume salesman." But Jim quickly learned the NC business and the advantages of COMPACT II, and together, Van and Jim (and later, Dick Weeden) expanded MDSI's reach across Texas.

Van asked Jim to please be sure to use the clients' names during the demos. Says Van, "The SE had usually been in contact with the prospect over the phone and perhaps in person, but I was usually meeting the customers for the first time at the demo, and I didn't have their names memorized." Van also preferred to walk through the plant or shop to see the customer's facilities and the types of products they manufactured. "This always gave me a better understanding of how we could help them."

Van and Jim noticed that quite a few people from the MDSI home office in Michigan wanted to check on their work in Texas during the winter months. During one such visit from Ken Stephanz, Van and Jim brought Ken to a call at Baker Oil Tool. When the presentation was over and they returned to the car, they talked about what went well and what could be improved upon. Ken said to Van, "You were chewing gum through the whole demo. You probably should have swallowed it."

Van knew Ken was right, and he was embarrassed. He vowed to himself that Ken would never have to mention *that* again. In fact, Van gave up gum chewing for good. (His group of friends today all know this story, so they often tease him by trying to offer him a stick of gum, knowing full well that Van hasn't chewed gum in fifty years.)

Van spent much of his time in Houston, and he remembers driving the 250 miles back to his home in Bedford (outside Fort Worth) many Friday afternoons, going 75 miles per hour on Interstate-45. To save time, he would rough out his trip reports on a piece of paper on top of his briefcase where it lay on the front seat beside him. "I know Judy loved trying to make out my scribbles," Van remembers. "I probably had a little Dutch mixed in once in a while."

Houston rapidly became MDSI's biggest hub of customers, including Cameron Iron Works, National Supply, Baker Oil Tool, Gray Tool Co., and Schlumberger. The importance of the oil industry to MDSI's bottom line was confirmed when Ken asked Chuck's team to add SIC codes (Standard Industry Classification codes) to the customer database. When they ran a report of the distribution of sales by SIC code, Chuck remembers, "the biggest percent of our business was in oil well tools." Unfortunately, this was just as the energy crisis of 1973 was causing fuel shortages and an economic downturn. It was a reminder of the importance of diversifying their customer base. But clearly Van and Jim Spencer had maximized opportunities for MDSI in the Lone Star State.

Van returned to Ann Arbor in 1975 and started MDSI's quality assurance department (QA). QA's responsibility was to review the thousands of machine tool links to see if they met the required specs for that particular tool.

By that time, MDSI's strategies for managing customer interactions had become more structured. For example, software bugs reported by customers or by field AEs were documented on "software problem reports" (SPRs) and forwarded to the link engineering group, headed at that point in time by Bob Dills (#183). When the problem was fixed, Van's QA group checked to make sure it really was. Gradually and of necessity, MDSI was becoming a complex, layered corporation.

And it had also gone international.

14

Flags of the World

In 1971, with MDSI's revenues growing steadily, Ken and Chuck made a trip to Europe to look at possibilities for international expansion. They attended an international machine tool show in Italy and then visited machine tool shops in England and France. What they found reminded them of their experience in the United States: a lot of interest in numerical control, but cumbersome computer languages, and a clear desire among part programmers for a faster and easier system. Ken remembers, "Everything that was available was just so inferior to what Chuck had designed and developed."

Using the Financial Projections software, Ken ran some scenarios for adding European customers and showed them to the MDSI board. "They were astounded with the numbers," Ken remembers. "They said, 'Absolutely.' By that point, our board was astounded every month by our financial reports, so they were already sold on the potential for further growth."

It was a big leap forward in the careers of both Ken and Chuck. Although Ken had traveled to Europe a couple of times when he was working at Bendix, that was just to attend a meeting, not to develop international business contracts. And Chuck had no experience working outside the U.S.

Ken remembers, "We hired local sales managers in each country and several engineers to work with them, and then we brought them back to Ann Arbor for intensive training." They came for six to twelve weeks to learn COMPACT II, get sales training from Mike Long, and shadow successful salesmen in the field. As these managers returned to their home countries and began signing up new COMPACT II users in machine shops throughout Europe, the part programmers were able to connect to timesharing computers via hard-wired networking between Europe and the United States.

. . .

Ken and Mike worked together to hire their first international manager, George Alexander (#181). He would head operations in the United Kingdom. Getting George was a real coup; MDSI lured him away from Rolls-Royce, where he was an expert in numerical control for the manufacture of jet engines. He turned out to be just what MDSI needed to push the company into the European market. Mike had great respect for George, and they became lifelong friends.

The first choice to lead the MDSI office in France wasn't similarly successful, and about six months in, Ken and Mike flew to Paris for a face-to-face meeting with the do-nothing manager. Mike remembers, "At the conclusion of the meeting, while the guy was still sitting there in the room with us, Ken turned to me and asked, 'What should we do?' I said, 'I think he needs to go.' So he was terminated." They soon replaced him with an ex-IBM engineer, John Adams (#231). "John had an English father and a Swedish mother," says Mike, "and he spoke good French, and he made the French operation go." John Adams also became one of Mike's good friends over the years.

By 1973, MDSI also had an operation in Canada, and over the next two years, the company added wholly owned subsidiaries in Switzerland and Germany. Mike remembers interviewing two really strong candidates to lead the Germany office. "We brought them both to Ann Arbor for interviews, and when we got together to compare notes, we all thought both were very strong managers. Ken said, 'Why don't we hire both of them?' So we did." Mike split Germany into two territories, put a manager in charge of each, and soon they had another successful European subsidiary.

. . .

Ken and Chuck, and Mike all understood the importance of keeping these new, far-flung locales connected with the home office. Mike still recalls the challenging start to one trip to George Alexander's MDSI office in Birmingham, England, with several other MDSI employees:

HOT TECH COLD STEEL

"We flew from Detroit to New York and landed in a driving rainstorm. We were to transfer to Newark for a British Airways flight to Heathrow. Starting out from the airport in two cabs, we got on the freeway, and shortly found ourselves stopped by a flooded underpass. We got out of the cabs and began walking along the freeway, carrying our luggage and cartons of handouts for the meeting. At an unflooded on-ramp, we hailed two more cabs, loaded up, and made it to Newark. Fortunately, the storm had delayed the British Airways flight, so we made the plane. We got to London late, but still had time to catch a train to Birmingham and conduct the meeting on time."

Mike traveled often to MDSI's international offices until the company hired Steve Imredy (#657) in 1977 as president of the international division. And many other MDSI employees in the United States regularly traveled to Europe to meet with their counterparts across the pond. For most of these men, international travel was a new experience.

When Bob Drewry and Dick Fisher (#132) traveled to a 1974 machine tool show in London, they were already old friends from their days

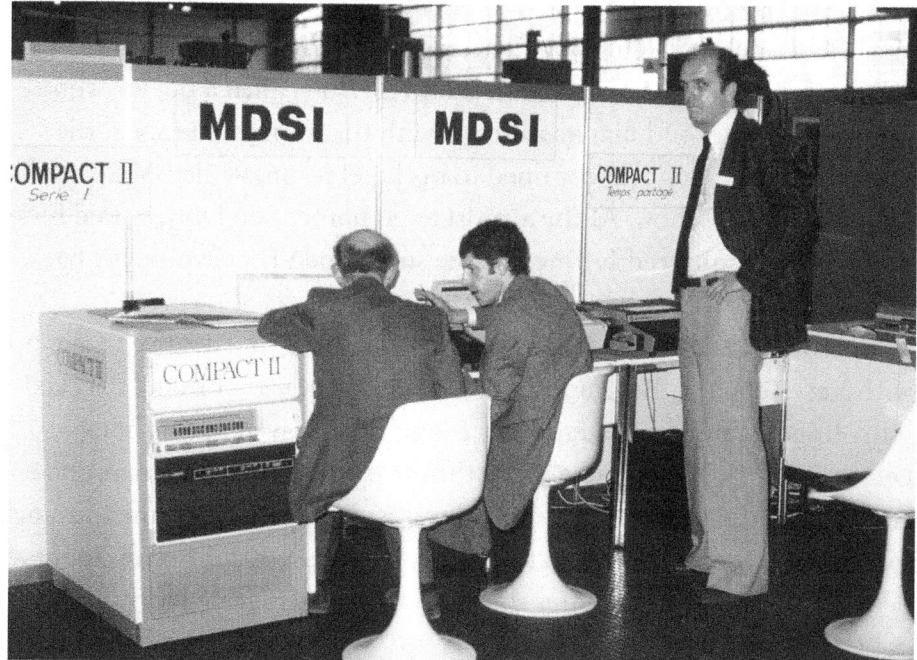

MDSI–France hosting a booth at "EMO" (the European tradeshow for manufacturing industries), 1978. Dominique Maitre (#628) is in the middle, with Pierre Bodereau (#237) on the right.
Photo: Urbanes Van Bemden.

working at Boeing in Wichita. As they waited in customs at the London airport, Bob took a look at their two passports. Bob's showed he was born in Snowball, Arkansas. Dick's showed he was born in Sugarland, Texas. "Here we were arriving in what could be called the capital of the world," Bob remembers. "Two country boys had made it to the big city."

They stayed in a nice London hotel. Bob had learned from Ken that if guests left their shoes outside the hotel room door before going to bed, the hotel staff would clean and polish the shoes before the next morning. After dinner that night, Bob told Dick about this service. Dick didn't believe him at first; he thought Bob was trying to play a trick on him. But Bob assured him it was a fact, and they said good night.

The next morning, as Bob was walking to Dick's room, he noticed a pair of shoes, all nice and shiny, sitting on a windowsill a short distance down the hall from Dick's door. Bob thought they looked like the shoes Dick had been wearing. When Bob knocked on Dick's door, Dick immediately opened it and said in an excited voice, "Some SOB stole my shoes!"

Bob retrieved the shoes from the windowsill, figuring some prankster had moved them on a lark.

Dick was relieved, but soon he wouldn't need his shoes. On their first full day in London, he started to feel unwell. Bob called a doctor who came to the hotel and diagnosed Dick with the mumps. He spent the rest of the week-long trip confined to his hotel room, while Bob worked the machine tool show. At the airport to go home, Bob Drewry told his friend, for the hundredth time, that he sure hoped Dick would get back to London someday.

Bob went back several times. On one trip in the mid-1970s, he met up with Ken Stephanz. Bob remembers, "I arrived ahead of Ken. I checked into our hotel and was taken to my room. I had stayed in this hotel before, but I had never had a room this nice. I lay down for a nap, and a couple hours later, the phone rang. It was Ken, and he asked me to come to his room. I was puzzled that his room number was four or five floors below mine; usually the better rooms were on the higher floors. When I got to his room, I saw there was a mix up."

As soon as Ken welcomed him inside, Bob said, "I think there's been a mistake. I think I have your room and you have mine."

Ken raised his eyebrows and said, "Let's go look."

On the way back up in the elevator, Bob assured Ken that he would be glad to switch rooms. Bob certainly wanted his boss, the president of the company, to enjoy the nicer accommodations. Ken wouldn't dream of kicking Bob out of his room, but he took one look around, picked up the phone, and told the front desk that he wanted a room like Bob's. Ken kept wondering aloud, teasing Bob, "How did you *do* that?"

The next evening, they caught a train to Leeds, north of London, where they had an appointment the following morning. When the train pulled into the station about midnight, they saw that the power was out in the entire city. It was pitch black. But the pay phone at the train station was still working, so Bob called the Holiday Inn, where they had reservations, to ask if they were still operating. He learned that the hotel would be able to show them to their rooms by candlelight, and, of course, there was no elevator service. After checking in, with a bellhop carrying a candle, Bob and Ken wrangled their heavy luggage up four flights to their rooms.

Bob remembers, "When we reached our floor, the bellhop unlocked a door for Ken and left him a lighted candle. Then the bellhop accompanied me down a couple of doors and let me into my room. I immediately went back to Ken's room to coordinate our schedule for the next day. Before I reached his room, I could hear him saying, 'I don't think this room is ready to be occupied.' I looked in and saw Ken on the phone, standing in the middle of a room that was full of ladders, paint cans, buckets, and no bed. Ken soon followed me back to my room, and it was, of course, pristine. Ken accused me of sabotaging him again."

The hotel quickly moved Ken into a nice room, but for the rest of the trip, Ken kept jabbing at Bob: "How did you *do* that?"

Ken liked to eat in fine restaurants when the opportunity allowed, and he was pleased to share his enjoyment of good food with his employees. When Bob and Ken were working a machine tool show in Paris for a week, they stayed at the International Hotel, and each evening, Ken would consult with the concierge to select a 4- or 5-star restaurant for dinner. After a few nights of this routine, Bob put his foot down. The next evening, when Ken met Bob in the lobby, ready for another night of gastronomic delights, Bob said, "I found the 'American Grill' here in the hotel, and I'm hungry for something more familiar."

Ken didn't hesitate. "That sounds good to me." Still dressed in suits and ties, they climbed onto stools at the bar and relished a couple orders of burgers and fries.

■ ■ ■

One of Ken's favorite memories of international travel is a memory of the home office. He was in Europe, calling back to Ann Arbor to get some information. Since he'd been away, Betty Ruddy had hired a new receptionist whom Ken hadn't met. When the young woman answered the phone, Ken asked, "Can I speak to Betty Ruddy, please? This is Ken Stephanz."

The receptionist said, "Oh, I'm sorry. I can't put you through. She left orders not to be disturbed."

Ken said, "That's okay. I'm her boss. She won't mind."

The receptionist was firm. "I'm sorry, sir, but she's not available."

Ken was flummoxed. "I'm the head of MDSI. Please put Betty on."

"I'm sorry, sir, I can't do that."

Ken never did get through to Betty that day. He always wondered what fear Betty must have instilled in that poor receptionist to keep phone calls at bay.

By the mid-1970s, employees of MDSI, no matter where they worked, were able to communicate via computer, thanks to Ron Peterson (#241), a former Comshare programmer. Hired in 1974 to oversee MDSI's fledgling management information systems, Ron was soon asked to create an internal, computerized communication system. The result was an email system that, Ron says, would "rival any email program you'd find today." It made it possible for MDSI to work around the world.

Any employee on the system could send a communication to his manager, or a manager could send a note to an individual employee, or to all employees in a particular department, or to all employees in an office. The system even included a return receipt request function. This strategy not only ensured that Ken could reach Betty no matter who was covering the switchboard, but that all the international offices could communicate with the home office at a time convenient for the sender and for the recipient.

■ ■ ■

Van also spent time in Europe, helping to get company operations up and running in England, France, Germany, Belgium, Luxembourg, Netherlands, Switzerland, and Scandinavia. He worked regularly with John Adams, the managing director of the Paris office.

By then, Van's oldest son, Jeff, was a young teenager, and Van taught him how to communicate with his dad overseas using the ASR33 that was installed in Van's house. (Jeff even attended a COMPACT II class when he was just thirteen years old, as an easy introduction to computer programming.)

Van would eventually transfer over to the international division full time as director of technical services under Steve Imredy. "My job was to see that each of our international subsidiaries had the right tools, training, and support to do their job," Van remembers. "The big question was, 'How can we, back at corporate, help you do a better job?'"

This was several years after Van had opened up the Texas market and then created the quality assurance department, so he was well versed in many facets of MDSI's operations. To the folks in the international offices, Van-from-the-home-office (whom they knew had been with MDSI since Day One) surely had the answers to all their questions. Van knew that wasn't true, but whenever he traveled internationally, he faced very long days. He remembers, "They would meet me for breakfast and practically tuck me into bed at night."

MDSI employees in Ann Arbor would sometimes express envy at Van's trips overseas, and he did enjoy himself, but he also admitted it wasn't all it was cracked up to be. "On the first trip to a new office, someone would meet me at the airport, take me to the hotel, maybe take me out to dinner and show me around. But after a couple of visits, they'd say, 'You know your way around. We'll see you at the office

MDSI management meeting in West Berlin, 1981, with MDSI staff from U.S., European, and Japanese offices. From left: Tony O'Toole (#215), Hiriyuki Nagasaki, unknown, Juergen Selig (#370), Urbanes Van Bemden, Olof Silfverling (#1175), Pierre Hertach (#1257), Dave Burnett (#761), Masaya Fukushima (#1015) and John Adams. Photo: Urbanes Van Bemden.

Monday morning.' I grew a lot professionally, personally, and culturally, but those trips meant a lot of time away from home."

This was especially problematic on June 24, 1976, "the saddest day in my life," Van says. He had flown that morning from Detroit to Milwaukee for a customer visit, but as soon as he arrived at Kearney & Trecker, he got a message to call the home office and talk to Betty Ruddy. Betty told Van that his youngest son, ten-year-old Gary, had been in a bad accident. There was a plane waiting for Van to return to Detroit.

"All the way back to Detroit Metro Airport," Van remembers, "I was sitting there wondering what kind of accident and how severe." An elder from Van's church picked him up at the airport and drove him to U-M Children's Hospital, explaining that Gary had been riding his bicycle on Plymouth Road, on the way to Daily Vacation Bible School, when he was hit by a car. "When I got to the hospital, I saw Brenda, and she told me that Gary had passed away." The tragic news rippled out into Van's circle of friends and coworkers, and many MDSI employees attended Gary's funeral.

■   ■   ■

With everything working out in Europe, Ken was ready to expand to Japan, the most important manufacturing market in Asia at that time. But Japan presented multiple unique challenges, the most significant being that Japan forbade foreign ownership of technology companies. In the mid-1970s, Japan's Ministry of International Trade and Industry (MITI) had granted only one exception to this rule—to Searle Pharmaceutical for the manufacture of its gelatin capsule.

When Ken first asked his attorney about setting up a wholly owned subsidiary in Japan, the attorney explained that this would be impossible. He said that a company with zero technology could be wholly owned by a foreign entity, and a medium-technology company could be 49 percent owned by a foreign entity, but a high-technology company like MDSI could have no ownership at all. In other words, MDSI couldn't reach the Japanese market unless a Japanese-owned company ran the show—and took the lion's share of the profits.

Ken was undeterred. He flew to Japan and met with a Japanese attorney and a Japanese CPA, looking for some sort of loophole.

"You can't do it," they told him. "MITI will never allow it. You can make an application to MITI for an exception, but they will turn you down. They turn everybody down."

Still undeterred, Ken sought an appointment with an official at the Ministry. At his first meeting, Ken brought along an interpreter, and he was glad he did. The official didn't speak a word of English. But, through the interpreter, the man listened patiently to the MDSI story and Ken's arguments for how MDSI's services could advance the manufacturing economy in Japan. The man nodded, thanked Ken, and sent him on his way.

One month later, Ken returned. He was received again by the same MITI official. Again, the interpreter facilitated a conversation. Ken felt welcomed. The official was very polite. Their conversation was interesting for both men. But still no hint of approval.

One month later, Ken was back. And again, one month after that. With each visit, he pointed out the benefits to Japan's manufacturing center and answered the official's questions. He sought to allay each stated concern.

After about five months, the official said, as interpreted to Ken, "I'm surprised. You're doing this the Japanese way."

Ken nodded. "Well, we want to work in Japan. Obviously, we should do it the Japanese way." To Ken, that meant solving every possible problem, addressing every possible argument in advance before any decisions were made. He knew it was important that nobody lose face in these negotiations.

After about nine months of monthly meetings, the official said, "Well, it's time for you to make your official application."

Ken did so, and then kept coming back every month for another six months. All that time, and all those trips, he was just hoping that MDSI would be able to do business in Japan.

At their fifteenth meeting, as the interpreter was speaking, the MITI official stood up, motioned to the interpreter to stop talking, and, in perfect English, said, "Mr. Stephanz, I am so pleased to tell you that MDSI has been approved. You have your exception."

Ken nearly fell off his chair. Not only would MDSI get to open a wholly owned subsidiary in Japan, but all this time, the MITI official had understood every word he'd said in English to the interpreter!

MDSI was the second-ever company to receive an exception from the Ministry of International Trade and Industry. One could argue that the services of MDSI in Japan did a lot more for that country's economy than a gelatin capsule ever would.

■   ■   ■

In May 1975, Ken hired Masataka Uesono (#367) to be the managing director of MDSI–Japan. "He was steeped in the old traditions," Ken remembers. "He'd bow and click his heels." One day as Ken was having lunch in Tokyo with Masataka and the Japanese CPA and attorney who were working with them, the foursome realized they had all been in the war between their two countries. Ken told them about his service in the United States Navy, while the CPA talked about the Japanese air corps and the attorney talked about his service in the Japanese army. Masataka had been in the naval college, training to operate the two-man kamikaze submarines (suicide torpedoes), called *Kaiten*, that the Japanese navy developed near the end of the war. Masataka looked at Ken and said, "So sorry."

Ken remembers how his work in Japan healed his own heart. "Having been in World War II, I didn't have a very high opinion of the Japanese when I started this effort. But I soon found that the Japanese people, in their own country, were the friendliest, happiest, and most honest people I'd met anywhere in the world. And I had traveled to a lot of other places by then. It was a very pleasant surprise."

Masataka set up an office in the Shinjuku area of Tokyo and hired two Japanese AEs—Yoshitaka "Yoshi" Taguchi (#397) and Seiji Kusaka (#405)—who then came to Ann Arbor for four months of training. Although they spoke some English, everyone quickly realized the language barrier was getting in the way, so Yoshi and Seiji spent their first two weeks in Michigan at a Berlitz language class in Farmington Hills. Van later learned that Masataka told his two employees that if they left their jobs before three years, they'd have to pay back to MDSI–Japan the $1,500 cost of the Berlitz class.

■ ■ ■

Yoshi and Seiji were still in Ann Arbor when a new MDSI employee from Frankfurt, Germany, arrived in Michigan for AE training. Alfred Vieth (#431) had just quit his job as supervisor of NC programming at Leitz (maker of the Leica camera) in Wetzlar, Germany, after three years there. Japanese competition in the camera market had forced Leitz to lay off a third of its 6,000-person workforce, and Vieth—with a wife and one-year-old child—was worried for his future there.

"It was a quite interesting and challenging time at Leitz," he remembers, "because they were introducing NC technology into their factories." At the time Vieth left, the plant had four NC lathes and five NC milling machines. Programming was being done manually and required four full-time NC programmers. "My job in the last year there was to introduce computer-assisted NC programming. For the lathe, we had H100 from Index, which was running on an IBM mainframe. It was written in Fortran IV, and I revised some of the code. For the drill and milling programming, I had developed my own system, written in Basic, which was a simple NC editor with plotting capabilities running on a Hewlett-Packard desk computer."

In 1974, Alfred had taken an NC control programming course at Siemens Erlangen and heard, for the first time, about a programming language called COMPACT II. He heard it was being offered in Europe by a Swiss subsidiary of an American company called MDSI. Shortly after that, he saw an ad in a German engineering newspaper (*VDI nachrichten*) about MDSI–Frankfurt. The newly opened office was looking for application engineers.

"One part of the MDSI advertisement caught my special attention," Alfred remembers. "The offer of a three-month training course in the U.S. I thought, *No matter how my start in this company will turn out, the three months will be a good investment and a gain of skills.*"

Alfred applied and was invited for an interview at the MDSI office in Frankfurt. By then, there were two managers on board, Peter Kopecky (#369) and Juergen Selig (#370); a secretary, Gabrielle Erk (#406); and one application engineer, Gerd Streckfuss (#392), who had been hired in September 1975 and was already in Ann Arbor for training.

Bob Drewry was in Germany to conduct technical interviews with AE applicants. "From the first minutes when we met," Alfred remembers, "we had such a good discussion on the MDSI concept and NC technology and also a lot of laughs, which made me feel very comfortable. Bob was my initial contact to MDSI's superior technical team, and I was eager to join and work for this company."

When Gerd Streckfuss returned to Frankfurt just before Christmas, his report to his new AE colleague, Alfred Vieth, was short and rather bleak: very long working hours (including Sundays), very cold weather, and not too much to see and do outside. But Gerd also gave Alfred a helpful hint. "Get in contact first thing with the two Japanese AEs in training, Seiji Kusaka and Yoshitaka Taguchi. They're staying at the same hotel, the Campus Inn, and they have a company car, so they might be able to get you from the hotel to the main office in the morning."

It was a cold and windy Sunday morning on January 4, 1976, when a twenty-seven-year-old Alfred Vieth said goodbye to his wife and baby son at the Frankfurt airport and took his first flight across the Atlantic. He was excited but also worried. Worried if he'd made the right decision to join an American company, worried about being away from his family for three months, and worried about the language barrier. He used his time on the plane to read MDSI sales brochures and brush up on the English he'd learned at a crash course in Germany.

Alfred landed at Detroit Metro Airport at 9:00 p.m. under a clear Michigan sky. Dave Price (#194) picked him up and drove him to the Campus Inn in downtown Ann Arbor. MDSI had made Alfred a reservation, and there was a welcome letter waiting for him at the reception desk. By the time he got to his room, his wristwatch—still on German time—said four o'clock in the morning. All he wanted was sleep. But just after changing into his pajamas, he heard two people talking in a strange language out in the hallway. He opened his hotel room door to find two young, Japanese men.

In English, Alfred asked, "Are you with MDSI?"

Yoshi and Seiji smiled and nodded and confirmed they were. After introductions all around, they asked if Alfred wanted to join them for a late-night trip to a local pizza place. His exhaustion forgotten in the excitement of meeting his new colleagues, Alfred put his clothes back on and followed them to their car.

The three men all piled into the front seat together and started heading south on State Street. Suddenly, they heard a police siren and people shouting on the sidewalk as a young man ran by their car. *Was he naked?* Alfred thought. The guy had been running too fast in the dark for Alfred to be sure.

But Yoshi and Seiji had been in Ann Arbor long enough to know just what was going on. They excitedly tried to explain to Alfred that the police were chasing a "streaker." It was a word—and a behavior—Alfred had never heard about before.

"This 'sport' was totally new to me," Alfred remembers, "and only strengthened my European prejudice that Americans (at least a few) must be 'crazy.' But after this surprise, Yoshi and Seiji and I had pizza and a few beers. Not too bad for my first day in the U.S."

MDSI turned out to be everything Alfred Vieth had hoped; his worries were for naught. MDSI–Germany was a commercial success, and he learned so much about all facets of numerical control—machines, programming, computers, software, and more. Plus, he now had colleagues all over the world.

After three years as an application engineer in Frankfurt, Alfred ended up moving to Ann Arbor for a few years to work for Jim Oyer (#302) in the link engineering department. Says Vieth, "It was one of the best working environments of my career."

■ ■ ■

When Yoshi and Seiji completed their MDSI sales training in Ann Arbor, Van returned with them to Japan to assist in their first demos. Van didn't speak Japanese, so he depended on his new colleagues to translate for him. At one demo with a potential customer, he watched as Masataka and Yoshi carried on a long conversation with a large group of personnel. "They seemed to be talking back and forth without getting anywhere," Van remembers. "So I asked Yoshi to ask them a particular

Van (left) with Japanese colleagues Masaya Fukushima, Hiriyuki Nagasaki, and Masataka Uesono (managing director of MDSI–Japan), congratulating each other on another successful year, Tokyo, circa 1982. Photo: Urbanes Van Bemden.

HOT TECH COLD STEEL

question. At that, a young man (probably a computer geek), said, in almost perfect English, 'That a good question.' Everybody laughed."

At another demo, the MDSI team was invited to stay for lunch at the company cafeteria. Van looked in dismay at his bowl of soup; it seemed to be floating with fish bones and squid. After he and Yoshi left the factory and were back on the train for the office, Van said, "In the future, if a customer wants us to join him for lunch in the cafeteria, please tell him we have to go."

Van was an avid photographer, and he remembers buying much of his camera gear in Tokyo's "Electronic Alley" near the MDSI–Japan office. He would eventually take more than 10,000 photos during his overseas travels for the company, many of which were published in an in-house MDSI magazine.

As Mike Long remembers it, MDSI–Japan was never the great success they had hoped it would be. A primary problem was that COMPACT II was an English-language program; it was never translated into any other languages. While the German and French customers were able to work with the English-language COMPACT II in a satisfactory manner, that wasn't the case in Japan. Mike traveled to Japan many times to try to push things forward—he figures he spent a total of four or five months there over several years—and he relocated a senior application engineer from the U.S. to Japan for a temporary assignment. But that subsidiary, Mike believes, was only ever marginally profitable.

Even so, it did expand MDSI's reach into new realms. As one example, MDSI–Japan introduced COMPACT II to Hindustan Aeronautics Limited (HAL), a state-owned aerospace and defense company headquartered in Bangalore, India. HAL had bought several Japanese NC machines made by Osaka Kiko, and they needed a good NC tape preparation system. When they asked Tsubakimoto Chain (a Japanese automotive supplier) to find out what system they should get, the immediate recommendation was COMPACT II. Yoshi Taguchi went to India to install a Nova (Series I)-based COMPACT II system and stayed on to teach four HAL engineers how to use it. For that multinational project, everybody communicated in English.

A gathering of MDSI's application engineers from around the world, 1978, left to right: Front row (fully seated on floor): John Adams, Pierre Bodereau, G. Howard Barrett, Dirk Van Krimpen, Bernard Auer, Gerd Streckfuss. Second row: Reinhard Hardtke (kneeling), Peter Kopecky, Gerrard Dawson, Andrew Thornton, Stephen Imredy, Carolyn Benivegna, Herbert Fehrensen, John Sinclair, Juergen Selig, Keith Schofield. Third row: Robert Neuerbourg, Alex Wilder, Juan-Jose Molina, Dominique Maitre, Urbanes Van Bemden, George Alexander, Christopher Jones, Masataka Uesono, Niels Goettsch. Back row: Eberhard Koehler, Harry Rock, Philip Markham, Anthony O'Toole, Gunther Schroedel, Roger Bianchi, Jean-Claude Morens, Alfred Vieth, Michael Smith. Not available: Seiji Kusaka, Norimitsu Shimizu, Yoshitaka Taguchi, Wilhelm Virnich.

HOT TECH COLD STEEL

● ■ ■

Throughout its international expansion, MDSI strove to always be culturally sensitive, though there were a few mishaps. Gerald "Gerry" Gorecki (#356) remembers setting up an exhibition booth for MDSI in Hanover, Germany. He thought it would be a nice touch to display the flag of every country where MDSI worked. He used poster board and colored pens to draw the flags for the United Kingdom, Germany, France, Japan, and so forth. He drew them from memory, but as he stepped back to examine the results, he thought it looked pretty good.

As the show opened, he noticed that Japanese attendees kept making a face when they looked at the display. "I finally realized I'd drawn the Imperial Japanese Army 'Rising Sun' battle flag, not the country's flag," Gerry remembers. "That didn't go over too big."

He could perhaps be cut some slack, if anyone had looked at his travel schedule. "For a time, I was traveling almost every day," says Gerry, "all over the United States, and sometimes to Europe. My wife was really getting tired of it, and so was I. At one point, I had walking pneumonia, but I had to get on a plane to Connecticut, so I had a prescription called in to a pharmacy in Hartford. I flew there, picked up my pills, and went about my job. That night on the phone, I admitted to my wife, 'I can't keep doing this.'" It was nearing the end of the year, and Gerry decided he would quit come January.

Soon after he was back in Ann Arbor, it was time for the annual MDSI Christmas party. As usual, MDSI would be handing out envelopes to all the employees with a financial gift from the company. Gerry remembers that his boss at the time, Rex Wolf (#525), pulled him aside, handed him the envelope, and then said, "Because you worked so hard this year, because you've been so loyal, here's a second one."

Gerry thought, *Oh, crap. Now I can't quit!* But they did get him some help to ease the burden. By that time, new employees were coming on board almost every day as MDSI was cornering the market around the world.

15

Computer Wizardry

With MDSI employees now operating across North America and Europe and in Japan, a shared corporate culture was more important than ever. With this in mind, Ken instituted an annual, three-day, company-wide meeting in Ann Arbor. Everyone flew in for these gatherings from wherever they worked. The event started on Thursday with each department meeting separately. These were important opportunities for folks who were doing similar jobs in very different locations to meet up and exchange problems, solutions, and novel ideas. On Friday, there were meetings across departments, creating unexpected synergies. And on Saturday, Ken gave a report to the entire group, summarizing where things stood, including any problems, and what new plans had come out of the previous days' conversations. As Ken remembers it, the close of his annual speech was always greeted with a standing ovation. "You cannot imagine the exciting ambiance of that whole situation."

The three-day gathering always concluded with a Saturday evening cocktail party and celebration in a hotel ballroom complete with hors d'oeuvres, a live band, and a dance floor. "It was unreal," Ken remembers. "Hundreds of people in one room creates quite a party." He's quick to clarify that it was always decent, never lewd. "But everybody had a wonderful time."

■ ■ ■

Ron Peterson, who ran MDSI's management information department (and developed the internal email system), remembers helping to prepare Ken's company presentations with daily, weekly, and monthly stats of customer growth. Says Ron, "I believe one major reason for the tremendous success of MDSI was Ken's uncanny ability to identify the important measurable units which would, in few words and mainly pictures, show the direction the company was moving. His penchant for identifying these variables allowed us to stay ahead of the curve when it came to hiring the right people in the right places to maximize customer service and support. It was a lot of fun at company meetings to show charts that often ran off the wall like the proverbial hockey stick."

A Chicago native, Ron learned he had a head for computing as early as 1960, when he prepared monthly reports using punch cards for the actuarial department at Kemper Insurance Company's Chicago headquarters. In 1968, he was going to school at Northeastern Illinois College, when he took his first class in Fortran and figured, "I can do this."

Looking for a job while going to college, he found Comshare's Chicago data center in the Yellow Pages, called for an interview, and was immediately hired as a nighttime operator. "My first day on the job I was assisting a fellow at Arthur Andersen with a Fortran problem using subroutines."

In January 1969, Comshare transferred Ron (and his new wife) to Ann Arbor. "I had no idea what I was getting into," Ron remembers. "I had heard about meetings at 'the apartment,' and I thought, *How nice, they have a corporate apartment for use by the employees.* I soon learned it was a restaurant and bar in the Huron Towers apartment building where the data center was located."

Ron's first big assignment for Comshare was to write a billing system to take all the usage data from the computers located in Ann Arbor, Chicago, and New Jersey, merge them together, and produce invoices and statements. The program was written in assembly language on the XDS 940 on terminals that had a screen and keyboard wired by serial ports to communications equipment that ran at a top speed of 300 bits per second, or 30 characters per second. By the time Ron completed that program, he was well aware of the usage data associated with Comshare's major customer, MDSI.

Ron soon took on the responsibility for all the business management programs at Comshare, including accounts payable, accounts receivable, payroll, and general ledger. Using Fortran, he developed and refined a business management program that ran on Comshare's XDS 940.

By 1973, MDSI's CFO Jon Ehrmann was pushing for MDSI to have a robust accounting and management system, so Ken offered to buy the Comshare system that Ron had developed. Ron remembers, "The two companies reached an agreement on price, and then Ken and Jon asked me to lunch with them at the Gandy Dancer—a restaurant in Ann Arbor's former train station. We talked about the opportunities of the two companies, and Ken made me the old proverbial offer I couldn't refuse. I never looked back or wondered if I had made the right decision."

As head of MDSI's management information systems (MIS), Ron immediately began to modify his programs to serve the needs of MDSI. "Since MDSI had employees in so many states and countries," Ron remembers, "one of the first challenges was to upgrade the payroll to accommodate all the craziness that state payroll taxation required. I used to believe that the delinquent sons-in-law of the governors of many states were chosen to write their payroll tax codes."

As with Comshare, Ron needed to gather computer usage data for MDSI, but that got more complicated when he had to account for transactions in France, England, Japan, and elsewhere. He says, "They used our software, but they had their own accounting and billing functions to bill their customers locally." Even so, those data had to be incorporated into the larger company reports. Whatever the MIS challenge, however, Ron always felt that Ken and Jon Ehrmann fully supported his team. "We had the resources; we had the computer time. Whatever tools we needed locally, whatever support staff we needed, we were provided it and allowed to grow."

■　■　■

With responsibility for all the data processing, Ron and his MIS team also had responsibility for data security, which included monitoring internal computer usage by MDSI staff members throughout the world. During one such routine audit, Ron found that the user accounts of the staff in England were being accessed after midnight there. Ron confirmed that the office staff in England didn't typically use the system after about 7:00 p.m. Further investigation revealed that the usage was, in fact, happening in California in the afternoons (when it was nighttime in England).

Over the next two months, Ron continued to investigate, with help from Tymshare and, eventually, the California Attorney General's office. When they finally identified the culprit, it was an application engineer who had left MDSI. He was working in California for one of his former MDSI customers and had somehow gotten or guessed the passwords of the MDSI users in England. By signing on as MDSI staff, he was saving his new employer tens of thousands of dollars in computer usage charges.

Ron remembers, "I was elated to be present in California when detectives from the Los Angeles Attorney General's office and local police officers, armed with a warrant, were able to catch the culprit in action." MDSI pressed for legal redress, rather than settling for a slap on the wrist, which confirmed (for Ron) that MDSI insisted on doing things right. The felon paid a settlement of $25,000 and forfeited MDSI stock he still owned. Plus, the case set a precedent when MDSI was reimbursed for all of Ron's expenses of time and travel required to catch the felon. Says Ron, "The case later showed up in legal textbooks and was studied by law students for a number of years."

Another breach of security occurred when some of MDSI's talented computer engineers hacked into the payroll system to satisfy their curiosity. It only took a couple of days for MIS staff to encode a bit-shift method of data storage and retrieval so that actual rates of pay and earnings were not discernable without knowing the algorithm. "It was a basic form of encryption," says Ron, "but it did the job."

■　■　■

In 1974, MDSI expanded its technical capabilities and services with two corporate acquisitions. The purchase of Hartec Corporation out of Pompano Beach, Florida, allowed MDSI to provide its customers a computer-based parts classification system called CODE. Today's readers may not remember when industries didn't have computer databases to track inventory, costs, and orders, but at the time CODE was created, this was a new capability. Hartec's coding and classification system helped manufacturers track parts, tools, raw materials, machines, and processes for improved management and cost savings. It was basically the first step toward just-in-time delivery.

CODE improved on the standard database design by using "graphic classification" of parts, allowing machine shops to organize their production according to the physical or manufacturing characteristics of parts. Sorting on one or more of the digits in the classification number allowed the user to determine the type and number of NC machines required to fill an order. Sorting in a different manner resulted in greater control of setup time, NC programming plans, machine loading, and so on.[25]

CODE had the potential to be a valuable addition to the machine shop's computerized capabilities, but as a product offer from MDSI, it never quite took off. Mike Long remembers, "We were not able to develop a business model for the CODE system that could generate revenue and lead to profitability."

More important to MDSI's future was the acquisition of a minor competitor out of Cleveland called NCCS-Word, Inc. The original Numerical Control and Computer Services (NCCS) was the company that Richard "Dick" Stitt founded after he copied SPLIT to create his own NC program called ACTION. Chuck had known Dick since 1959, while Chuck was still at Buhr; the two men had since pursued somewhat parallel paths.

After MDSI acquired his company, Dick Stitt (#283) was welcomed onto MDSI's executive team as vice president of operations. Mike Long immediately liked Stitt, and Ken soon considered his new VP a "fixer of all problems."

Bruce remembers Dick walking through the office introducing himself, when one member of the R&D team—an outspoken and rather aggressive personality—walked out of his office, blocked Dick's path, and demanded, "Who are you?"

Dick politely shook the man's hand and said, "My name is Dick Stitt. It rhymes with shit."

Dick Stitt (left) and Chuck Hutchins, 1974. Photo: Bruce Nourse.

The aggressor was taken aback by the self-deprecating humor and ducked back into his office just as everyone else burst out laughing.

Dick brought some of his NCCS team with him to MDSI, including his son, Doug Stitt (#301), an experienced programmer. Doug was folded into Ron Peterson's MIS group and, as Ron remembers, "quickly became the *force majeure* for everything we were able to accomplish thereafter. He blew me away with his coding skills. I couldn't even fathom how he came up with some of the solutions we needed as a company." These solutions included the inventory management and billing interface for the thousands of machine tool links that were leased on a monthly basis.

Ron continues, "Doug was so fast and so accurate in all he did that it kept me on top of my game. If I hadn't planned properly or communicated properly, within hours he would be correcting my mistakes. What I never quite got over was that Doug was the fastest person on a keyboard I'd ever seen, and he only used two fingers!"

Doug Stitt is still amazed at how much the MIS group accomplished using the Fortran language. "We developed all these subroutines for printing data in various formats. It was a perfectly valid language. We

HOT TECH COLD STEEL

could very quickly write these applications. Looking back now, it was kind of amazing what we did."

∎ ∎ ∎

By the time Dick Stitt's company merged into MDSI, NCCS-Word's latest version of NC software, ACTION II, worked on minicomputers, which was something MDSI was beginning to explore. Today's readers will consider "minicomputer" a misnomer, as these machines were still the size of a wardrobe. But compared to a room-sized mainframe with fifty platters, minicomputers were a revelation. One of the earliest minicomputers, the PDP-8, was a 12-bit machine from Digital Equipment Corporation (DEC), introduced as far back as 1965.

When MDSI was starting up, minicomputers held little interest for Chuck, as they could not compete with the memory of the XDS 940's 48 KB of memory. But in 1970, DEC introduced the PDP-11 with 64 KB, and soon after that, Texas Instruments revealed the TI-980 with 128 KB of memory. In fact, the TI-980 was a 64K word-addressable computer that offered twice as much memory as the PDP-11 and 2.66 times the memory of the XDS 940. Chuck was intrigued: "We were tired of fighting space problems. The idea of having twice the memory made your eyes bug." MDSI bought a TI-980 to explore the possibilities.

Chuck was further intrigued when he learned that a small competitor out of California, called UniAPT, was running NC software on a minicomputer. Chuck did a little research and discovered that UniAPT was using a program written for the 24-bit mainframe XDS 930 on the 12-bit PDP-8. Chuck considered how this might have been accomplished. He said to Bruce, "I wonder if they're putting 24-bit words in two 12-bit words on the PDP-8, with software to fetch and decode the 930 instructions." Chuck asked Bruce if he could write an emulator for the TI-980 to fetch and decode the XDS 940 instructions. "Don't worry about disk reading, etcetera," Chuck told him. "Let's just check plain code execution and see if it works."

In almost no time, Bruce created the necessary computer emulator software, and he also wrote an interrupt-oriented operating system (OS) in the image of the XDS 940. The TI-980 produced exactly what they expected. "Wow!" was all Chuck could say. They now had the potential to move COMPACT II to minicomputers without rewriting the assembly language software that had taken many man-years of development. (See photo of Bruce and the TI-980 on page 45.)

When a Texas Instruments field service engineer found out what Bruce had done, he said that MDSI was the first to implement an interrupt-oriented OS on the TI-980. That information spread through TI, and Bruce would sometimes get calls from TI engineers when they had an interrupt problem.

The 128 KB memory of the TI-980 was a bonus. Bruce's operating system needed 16 KB, and the XDS 940 emulator needed another 16 KB, but that left 96 KB of memory unused. That was enough memory to allow two typical timesharing users to run programs simultaneously on their own in-house minicomputer system.

Thus, MDSI saw an opportunity for additional revenue. At that time, a TI-980 cost around $15,000 (in 1970 dollars), but loaded with COMPACT II and installed in a machine tool shop—and rebranded as MDSI's "Action Central" system—the same machine could be sold for $100,000 (something like $750,000 today).[26] Despite the price tag, customers who were used to paying thousands of dollars a month in timesharing fees were immediately interested in having their own, in-house computer.

Seth Powsner was working at MDSI that summer, and he remembers that Chuck delayed shipment of any TI-980s until the tapes it generated were identical to the tapes produced by the XDS 940. Seth remembers Chuck always said, "Program so it works correctly, and don't stop until it's absolutely correct, every last bit."

Of course there was reluctance among MDSI's investors about selling minicomputers to customers that had previously been a source of ongoing timesharing revenue for the company. But as Mike Long pointed out, "I never argue with the marketplace." By 1981, MDSI had sold 300 Action Central units (later called the MDSI 400 series).[27]

■　■　■

　　　　　　　　　HOT TECH COLD STEEL

A few years into MDSI's phenomenal growth, Chuck turned to Seth Powsner with a serious problem. As Chuck remembers it, "Our computer disk storage was becoming overloaded. That meant we had to reduce the number of users assigned to a computer, and that would have cut into our profits. Today, we'd never worry about something like that, but disk storage limitation was still a big deal at that point."

Chuck had the idea to rewrite the Quick EDitor (QED) to store four 6-bit characters instead of three 8-bit characters in the 24-bit word. "On that basis," Chuck explains, "50 megabytes would become the equivalent of 66." It was an interesting idea in theory, but it meant a daunting rewrite of QED and reworking of COMPACT II's read routines. After Chuck told Seth what he had in mind, Seth figured out how to do it, solving all the new problems that this approach created down the line.

During the rewrite, Seth made the editor commands uniform; that is, every command option worked anywhere that it was logical. "We observed an incredible—and unexpected—gain in CPU efficiency," Chuck remembers. "This was a result of changing the conversion of character pointers to word pointers from using a divide-by-3 (a 9 CPU cycle instruction) to a right-shift-2 (a single cycle instruction), which did the equivalent of a divide-by-4. We couldn't believe the CPU gain we got from that. That upped our profit significantly." They called the new command SBQ (for Six Bit QED).

Seth remembers that "some of the gain came from reduced file IO." By the fall of 1974, he was working on an 8-bit retro-fit of SBQ to speed it up for the PDP-11 and TI-980. He was doing this work for MDSI out of his med school dorm room at Yale.

■ ■ ■

Chuck was always attending to the latest developments in computing technology and how these might be useful to MDSI, and his group of programmers pushed each other to adapt. "My whole R&D team was highly motivated to write the best code they could," says Chuck. "Though we joked that MDSI stood for 'Many Dumb Stupid Idiots,' we were just as likely to agree with Roy Winn when he said that it stood for 'Many Devoted Superior Individuals.' That's a good description of my guys."

Despite the breadth and depth of MDSI's incredibly talented programmers, Ken and Chuck never considered buying their own mainframes to provide computer timesharing directly to customers. First, they would have had to replicate the network of computer locations that were already set up by Comshare and Tymshare. But more importantly, the computer industry was rapidly changing, and they didn't want to end up saddled with obsolete technology.

Of course, by 1975, the XDS 940 was itself becoming obsolete technology. It wouldn't be long before Comshare and Tymshare would be facing difficult choices about upgrading. Plus, the latest computer languages being taught to the newest generations of programmers had little to do with COMPACT II, which was built in assembly code specific to the XDS 940. MDSI couldn't hire new programmers who knew it; they had to be trained before they could be productive.

In sum, the challenge of this era was how could MDSI prepare for coming changes when its entire framework of services—COMPACT II and the thousands of machine tool links—were developed for the XDS 940?

One answer came in the form of DEC's PDP-10 model KL10D computer, one of which was housed at an Ann Arbor company called Automatic Data Processing (ADP). Bruce remembers, "The DEC-10 had the capability to do microcoding. That meant you could take out the native instruction set of the DEC-10, and you could replace it with another entire instruction set of your own choosing. We wondered if we could take all this XDS 940 assembly language code and write microcode to make it the native instruction set of the DEC-10."

At that time, Digital Equipment Corporation, headquartered in Maynard, Massachusetts, was one of the largest computer companies in the world. They employed thousands of people, including many of the nation's leading programmers. The guys at DEC told Chuck that only two persons outside the company had ever microprogrammed the KL10D. Chuck remembers, "They gave us a snowball's chance in hell of doing it." Chuck told the DEC representatives that he had spoken about his idea with Dave Moon at MIT—who had written microcode for the KL10D—and Dave thought it was doable.

With that objection thwarted, DEC threw up another roadblock: the microcode compiler had a $40,000 ticket price. Chuck said, "The cost is just a number we'll use to evaluate the cost-benefit ratio for the project." There was no question that MDSI could afford it.

DEC's final roadblock was to warn Chuck that the KL10D received only "Level C" support from DEC.

Don Redding (#381) was with Chuck at that meeting, and he interrupted to say, "Chuck, Level C means that if you have a problem, DEC will light a candle and say a prayer for you."

Ignoring all these roadblocks, Chuck hired Dave Moon to consult on the project, and Moon created a prototype microcode that MDSI started using with the PDP-10 at ADP. Every job that ran through the PDP-10 ran twice as fast as it had on the XDS. Chuck's idea was working.

"A year of PDP-10 rental instead of two XDS 940 rentals saved us $600,000," Chuck remembers, "which went right to the bottom line!"

And then Dick Wagman got involved.

■　■　■

Dick Wagman (#351) came to MDSI in February 1975 right out of college, having spent most of his undergraduate time in the University of Michigan computing center working on IBM mainframes. As a young man, he'd seen enough of IBM's corporate culture to feel skeptical of all corporations, but when Chuck took a chance on him, Dick took a chance on MDSI. "It turned my attitude around," he says, "because I saw that we could do excellent work and make good things happen. At MDSI, I didn't have to feel like a sellout."

MDSI initially hired Dick because of his experience with IBM mainframes and the Michigan Terminal System (MTS), which was an IBM-based timesharing operating system. MTS was not commercially available, but some of MDSI's customers were using other IBM systems, and MDSI wanted to make it easy for those systems to interact with what MDSI was offering. Dick remembers, "IBM had a number of operating systems, ranging from pretty decent to appalling, that real customers actually used." One of Dick's goals was to figure out how to port code from one operating system to another. Some porting involved emulation (like Bruce had done when emulating the XDS 940 on the TI-980), and Dick was developing a general understanding of when to abandon emulating and "go native and start actually using the real code for that particular operating system or computer."

Dick continues, "Then along came the PDP-10 project, and they wanted to microcode the thing. I had always dreamed of doing microcode, but at that point, I hadn't even gotten near microcode; I didn't really know what it was. I had no clue about what its concepts were." When Dick Wagman got a look at Dave Moon's prototype, he wanted to know more. "Needless to say, I got my grubby little paws into his microcode. The way I worked stuff out was I'd take these lists and throw them on the floor of my office, take off my shoes, and study them until around three in the morning. If I kept doing that for a few nights, lo and behold, I started thinking I understood it." After some time examining the microcode for the KL10D, Dick told his colleagues, "I think I can do better than what Moon did. I think I can squeeze a fair amount of extra performance out of it."

Wagman, who was just twenty-five years old, understood that MDSI was beginning to hang the future success of the entire link-writing department on solving this microcode challenge. In that sense, the future of the company was riding on his work. The only way MDSI could remain competitive in the long run was if its thousands of links could be migrated to a faster, better computing system.

For months, Dick holed up in his office with a yellow legal pad, throwing lots of paper on the floor as he tried various approaches. When he finally had something he thought was worth testing, he needed access to a standalone PDP-10. ADP agreed to let Wagman come to their westside Ann Arbor facility after midnight to run some tests. There were bugs at first, of course, but he could see it was starting to work.

By the end of 1975, MDSI had hired Al Kortesoja (#423), a compiler expert who had some idea of what Dick was trying to do. After many late nights testing his code, Dick reported to Al, "I think it really works.

There's one instruction that I didn't test, but I'm confident." Al told him to go back the next night and test that one instruction.

Dick remembers, "I went back there, and I managed to rig a difficult test case that tried that micro-instruction. It worked. I came back and said, 'Yep, it's good.'"

Dick thought Al looked like he was going to have a heart attack at the thought of going live with Dick's code instead of Moon's prototype. But they had to try it.

Al said, "All right, you make sure you're in here at seven tomorrow morning, just in case anything happens."

But Dick was never able to get *anywhere* at 7:00 a.m. He remembers, "I rolled in around 9:30, and I poked my head into Al's office and said, 'Any issues? Everything okay?'" The whole place was calm and quiet. Customers had been using Dick's microcode, but nobody knew that anything was different. Dick thought, *Holy cow! It works.*

Suddenly, every job going through the PDP-10 was running 15 percent faster than before. Even more important, Dick's microcode ensured that all the machine tool links could be migrated to a new system. Says Dick, "That gave us a good intermediate product as Chuck and his wizards got it all interfaced. I left lots and lots of breadcrumbs in the program to help them figure out how to ultimately hook it up to any other system. I always hoped that if anybody wanted to, they wouldn't struggle with it."

Some weeks later, Al was talking with a senior executive at DEC who told Al, "You know, there are only about four people in this country who know how to microcode this computer, and you've got one." Dick Wagman became a "grade A microcoder" and eventually left MDSI for a job at DEC, where he made their own machine run 10 percent faster. (Another member of Chuck's R&D team eventually ended up at DEC as one of their top compiler architects.)

With Dick's improvements, MDSI rented a second PDP-10 from ADP, reducing further the rentals of XDS 940s and dropping $1.2 million in savings to the bottom line.

Ron Peterson, head of the MIS group, was tracking the huge gain in computing power per dollar spent on timesharing computers that MDSI didn't own. He remembers, "There was a long, and at times emotionally charged, study of whether to purchase equipment and bring it in house or look for a different vendor to supply what we needed." As it turned out, ADP in Ann Arbor was able to offer MDSI all the power needed to serve customers across the globe. By the end of the 1970s, MDSI made a successful switch away from Comshare's and Tymshare's 940s to ADP's newer and far more powerful PDP-10s (and, later, PDP-20s).

· · ·

Though the KL10D computer worked for the time being, the complications involved in porting COMPACT II from one computer to another called for a more systemic solution that didn't rely on the specific language of any one vendor's equipment. So in 1976, under the leadership of Al Kortesoja, the R&D group began writing all new software in the Pascal computer language. "Hindsight says we should have chosen C," says Chuck, "but C was still in its infancy in 1976."

Unlike the assembly code for the XDS 940, Pascal was a higher-level language that computer engineers were learning in school, so new hires at MDSI were now able to be productive as soon as they arrived. Pascal was also a generic language that would run efficiently and effectively on varied computers. Al Kortesoja remembers, "We wanted a compilation system that allowed the Pascal code to automatically translate to a variety of different computers, so that we could offer our product on whatever computer made the most sense for the marketplace." The Pascal compiler design included a separate Object Code Generator (OCG) that worked much like the machine tool links of the NC programming system: with any new computer, the programmer only needed to tweak the OCG and all the Pascal code would then run on that new computer (such as DEC's VAX computers, the successor to DEC's PDP line).

"We called the language, brilliantly, 'MDSI Pascal,'" Al jokes. "We actually had a contest to name the language. First prize was a one-week trip to Cleveland. Second prize was a two-week trip to Cleveland. The winning submission was 'Ask Al,' because that's what everybody did."

Al Kortesoja, part of MDSI's R&D team, at an MDSI summer party, sitting on the "dunk tank." Al remembers, "We had a tradition of putting managers in a dunk tank and letting the staff have a shot at dunking them. I've got my Pascal T-shirt on and an MDSI headband. The computer I'm holding is a disemboweled Data General MPT—which we referred to as a 'Muppet.' It was Data General's terrible attempt at a PC-type computer. Between me and the computer, everybody wanted to try their hand at the dunk tank." Photo courtesy of Al Kortesoja.

Kidding aside, the increased flexibility offered by MDSI Pascal encouraged Bruce Nourse and Bob Johnson (#475) and others to develop new products that MDSI might want to incorporate into a future business plan. One was called DESIGN, which stood for Design Engineering System Including Graphic Notation, and was an early form of three-dimensional volumetric CAD software entirely written in Pascal for the PDP-10. Al remembers, "It really had nothing to do with the NC product, but the intent was that, eventually, a user could go directly from a drawing to the machine tool, skipping the part programmer and teletype tape steps in between."

Another group of MDSI engineers used Pascal to begin writing a program to replace COMPACT II. They called their fledgling effort ANC, which stood for Advanced Numerical Control. Bruce remembers, "Those of us who didn't like ANC, me being one of them, would say that ANC stood for Ain't Never Coming." The naysayers were prescient: the ANC team worked on the program for several years, but not much came of it.

■   ■   ■

Rare missteps aside, MDSI's investment in R&D was clearly pushing forward technologies that went beyond numerical control. In fact, in 1976, Chuck's team introduced what was, for all intents and purposes, a personal computer for MDSI customers.

The "Smart Terminal 1" (ST-1) was built around an IMSAI 8080, an early microcomputer with 32 KB of memory. MDSI bought an IMSAI 8080 and built a cabinet around it to create a work surface on which to place the paper tape punch. Mike Levine, Chuck's friend whom he met in 1962 at Conductron, was co-founder with Ed Zimmer of Ann Arbor Terminals. It was one of their first cathode ray tubes (CRTs) and keyboards, model #K4080D, that was attached as an integral part of the ST-1 in the spring of 1976. Tom Zwitter (#292); John Whitney (#378), who had built an IMSAI 8080 from a kit while still in high school; and Jeff Broughton (#482), who had just graduated from MIT and would go on to become one of the country's foremost experts on supercomputers, began writing the software to make the IMSAI into an intelligent terminal that could communicate with MDSI's timesharing computers.

The purpose of the ST-1 was to reduce the connect time for customers by allowing them to print programs, prepare their edits, punch paper tapes, plot the tool path, and save to a floppy disk while off-line from the timesharing system. This saved customers money, and it freed up the timeshare computer to sell more connect time to other users. Thus, MDSI was still able to make its connect-time fees, plus the added revenue from selling the ST-1 machines.

Chuck's team set a goal of exhibiting the new ST-1 at the 1976 International Machine Tool Show in Chicago, which opened the Wednesday after Labor Day. Seth Powsner was working at MDSI that summer, and he still remembers Zwitter riding herd on that project. "His wife, Becky, would bring him dinner in the parking lot," Seth remembers.

The weekend before the Chicago show, the ST-1 still wasn't working quite right. Everyone was putting in long days and nights. Finally, it was time to ship the machine where it would be set up in the exhibition hall. The R&D team had it working by about 1:00 a.m., and Chuck said, "Okay! Pack it up! We're just going to make the flight."

But Seth said, "No, we've got to take this apart and put it back together at least twenty feet away to see if it will work when they set it back up in Chicago." Despite the late hour, nobody argued with that. So here they were, amongst the cubicles, taking the smart terminal apart, carrying the pieces across the room, and putting it back together. By then, it was 3:00 a.m., and the sales engineers were calling to say, "Where is it?"

Seth looked over and saw Chuck lying on the floor groaning, "I think I'm having chest pains."

Seth, who was by this time in med school, laughed and said, "Chuck, you're not getting out of this that easy."

The rebuilt machine worked as planned, and they got it on the plane to Chicago. As it turned out, though no one realized it at the time, MDSI introduced the ST-1 exactly two days after the Apple I computer, hand-built by Steve Wozniak and introduced for sale by Steve Jobs, was revealed on Labor Day 1976 at a computer show in Atlantic City, New Jersey. "But the Apple-1 had only 8 KB of memory," Chuck points out, "and no software. Ours had 32 KB of memory and came completely programmed to do all the tasks that an NC user needed to perform between computing the part program on the timesharing system and executing the control tape on the machine tool."

MDSI eventually sold more than 1,000 ST-1s.[28] The production models were built by Carterfone in Dallas, Texas.

To keep things simple, the ST-1 only had RS-232 interfaces, with DB-25 connectors for peripheral devices. These used 3 wires: send, receive, and ground. Some devices had the signal wires reversed. To make life easy for the user, a dual set of plugs for each IO device was provided, and one was wired for each polarity. If one was wrong, the other was right. The user could just try each plug until things worked. There was no need to worry about finding a send/receive swapping adapter; it was built in.

Furthermore, the ST-1 used the same editing commands as QED and SBQ, which were the text editors on the XDS 940.

■　■　■

Al Kortesoja was at MDSI the day some "highfalutin VP of a company like Unisys" stopped in for a visit. After the VP heard about all the things that Chuck's team was working on, he said, "Don't you guys just make control tapes for machine tools? Why in the heck are you doing all this stuff?"

As Al recalls it, Chuck merely responded, "Because we need to."

Says Al, "Chuck's genius went well beyond COMPACT II. Without a vision like Chuck's, MDSI would have hit a dead end. With that vision, we ended up where we wanted to go most of the time. Sometimes we didn't, but in any case, we learned a lot on the way, and that's never a complete waste. In fact, many people built their subsequent careers on all this wizardry we did at MDSI. They went on to other companies, like DEC and Lawrence Livermore, and their new colleagues asked, 'Where did you learn all this stuff? How did you figure this out?' The answer was MDSI."

16

Measures of Success

As MDSI's profits grew into the millions, Ken insisted on perks for his extremely hard-working senior staff, which included their choice of "company car." At one point, Ken had a Mercedes coupe while Mike Long drove a BMW. It was a bit extravagant, but Ken was looking for ways to reward his team for always going above and beyond.

The executive team eventually included Ken, Chuck, Mike, financial manager Jon Ehrmann, personnel manager Dennis Mummert (#404), Dick Stitt, Rex Wolf, John Shuman (#634), John Montjoy (#870), and the head of the international division, Steve Imredy. Once a year, this group decamped from Ann Arbor for an executive staff retreat. These multi-day gatherings were held most years at a resort in Hilton Head, South Carolina. The company's leaders immersed themselves in business matters until mid-afternoon, took some time off to golf or sail, and then reconvened for dinner and drinks. It was a combination of team building, problem solving, strategic planning, and vacation, though it was not at all restful. Mike Long remembers, "We would thrash out all the issues and come up with a plan for the coming year. Every one of those annual meetings was successful, because we always left with a plan that, for the most part, we executed successfully."

■　■　■

One significant strategic decision was to build a new company headquarters. After four years working at their Main Street location, MDSI's Ann Arbor team was clearly outgrowing its office space. But renting even more square footage felt like a waste of resources. They considered moving out of Ann Arbor, despite the connections to Comshare and the University of Michigan. After all, MDSI was now firmly established as an international company. The headquarters could be anywhere in the United States.

"A little-known fact," says Ken, "is a lot of us wanted to move south, not stay in the cold climate of Michigan." Ken put together a small scouting team to check out possible locations in North and South Carolina, Florida, and Texas. "But the one rule we established in the beginning was that, whatever location we proposed, everybody on the executive staff had to agree to it." With each new idea, everybody did agree, except Chuck Hutchins.

Finally, Ken pinned him down: "Chuck, would you move *anywhere* out of Ann Arbor?"

Chuck had to admit the truth: "No." Faced with the real possibility of leaving Michigan, disconnecting from his contacts at U-M, uprooting his family, and saying goodbye to the small but cosmopolitan community of Ann Arbor where he'd lived for twenty years, Chuck couldn't do it. He wanted to stay with MDSI, and he wanted to stay home.

So in 1974, MDSI bought fifty-two acres on Ann Arbor's Plymouth Road, just east of Earhart Road and the exit ramp off US-23. (The buildings are now easily found at the entrance to the Domino's Farms complex, but Domino's Farms wouldn't be built for another decade.) "We looked everywhere for the right property," Ken remembers, including the Research Park on South State Street and land on the western edge of the city. "This was a beautiful piece of land. It was not hemmed in by anything. It had room for expansion. It turned out to be the right place." Chuck saw an additional selling point: an easy three-step set of directions from Detroit Metropolitan Airport—1) go west on I-94, 2) turn north on US-23, and 3) turn right on Plymouth Road for about a mile to the MDSI headquarters on the left.

■   ■   ■

At first, the large piece of undeveloped land was rather daunting. Before an office building could be designed, MDSI needed an entrance, roadways, parking lots, greenspace, and a plan for construction that kept future options open. Ken turned to Carl Johnson of the Ann Arbor landscape architecture firm of Johnson, Johnson & Roy for the initial

A company softball game on the MDSI corporate campus. Judy Foster Leverett is at bat, with Terry Myers (#244) catching, and Dick Stitt as umpire. Photo: Urbanes Van Bemden.

site planning. This company had worked on campus master plans for the University of Michigan and the University of Wisconsin, so a relatively modest corporate office park was no problem.

The result was an especially pleasant, natural place to work. The buildings and pavement were kept near the entrance, so that behind the buildings, employees had access to swaths of green lawn, flower gardens, and picnic tables, with walkways that wound around a small pond. According to Seth Powsner, the only thing missing was a bicycle rack; he had to lock his bike to the natural gas meter. North of the building site was plenty of room for a softball field where employees would play many games of softball and where an annual company picnic became a new tradition. (MDSI bought an additional twenty-five acres in 1979, bringing the total size of the complex to seventy-seven acres.)

When it was time to start the first building, MDSI turned to a network of established companies. Ralph Bergsma, a longtime friend of Ken and of MDSI investor George Simon, took over project development through his construction consulting company, Property Development Group (PDG). PDG had developed the building at 320 North Main where MDSI was currently housed. PDG brought on the architect from that project, Warner & Summers, Inc., to design MDSI Building I. The contractor, Holder Construction of Atlanta, Georgia, was Dave Morgenthaler's recommendation.

The 40,000 square-foot building was completed in July 1976, but by the time it was ready for occupancy, it was already too small for the size of the company. Not all of MDSI's 160 local employees were able to move to Plymouth Road. Some stayed on at the Main Street offices, as planning began for a second building, twice as big as the first.[29]

When Building II opened in September 1979, MDSI had 416 employees in Ann Arbor (with another 150 working in the field around the U.S. and internationally). The new building was three stories with a basement. By July 1981, MDSI had added a third building to the site— another 80,000-square-foot, three-story structure, designed by Sigmund Blum & Associates of Bloomfield Hills.

All three buildings in the cohesive complex were steel-framed, brick structures. Plentiful, long windows maximized natural light, allowing for an abundance of potted plants. The windows looked out on a stately elm tree and a large, outdoor, aluminum sculpture commissioned by MDSI from sculptor James C. Myford. The three-part abstract piece was intended to represent metal shavings coming off a metal cutting machine. "It all turned out beautifully," Ken remembers. "It was like a Garden of Eden."

Building III included a glass-domed employee dining room with seating for 300.[30] The cafeteria—managed and operated by Stouffer's, the frozen-food giant—served lunch daily. This was a big timesaver, because in those days, there were no lunch spots out on Plymouth Road. Soon enough, in recognition of the needs of MDSI's many employees who were single and working long hours, Stouffer's began serving three meals a day. Says Chuck, "That's when our employees really got to know each other." Reflecting on that unusual offering, which is now common in tech companies, Ken has called it an early experiment in "social media." In fact, the frequent socializing in and out of the workplace led to a few marriages.

Employees moved amongst the three-building complex through elliptical, enclosed, transparent walkways that brought in sunlight

MDSI's international headquarters in Ann Arbor, Michigan, after all three buildings were completed, circa 1981.

One of the "gerbil tubes"—the walkways between MDSI's three buildings. Photo: G-Photographic, Oak Park, Michigan.

but kept out the cold. Chuck remembers that Ken told the architect, "I don't want these to be like the covered walkways at the airport. I want something dramatic!" Employees affectionately dubbed the Plexiglas hallways "gerbil tubes."

John Samford (#479) recalls one summer day when tornadoes were forecast while he was at work. "The sky turned green. We all went out to the tube to watch the weather, and we were chastised for putting ourselves at risk."

Samford's memories also reveal that symbols of corporate hierarchy were still practiced in the new location. While the buildings had ample offices with doors, cubicles were also part of the layout. Samford and other members of the link-writing team were assigned cubicles with fairly low walls. He says, "I'm very proud of how collegial the group

HOT TECH COLD STEEL

was, practicing what was then called 'egoless programming'—helping each other, sharing code, offering suggestions, etc. We were ensconced in individual cubicles, but there was a lot of over-the-wall chatter. Even so, we envied the programmers in R&D who had 'higher walls' on their cubes, giving the illusion of greater privacy."

Seth Powsner claims all the carpeting in the new buildings was a dark shade of brown, thanks to the practical analysis of his colleague Dick Stitt (who oversaw the building projects as VP for facilities and contracts). "When the carpet guy came in with his samples," Seth remembers, "Dick told him to lay all the brown samples out on the floor, then Dick took a large cup of coffee with cream and proceeded to pour it over all the samples. He told the carpet guy, 'Come back in an hour. Whichever one doesn't show the stain, that's the one we're taking.'"

Dirk Van Krimpen (#676) first saw the MDSI headquarters on January 2, 1978, when he came to Ann Arbor from Europe to train as a new application engineer. A former manufacturing engineer at the Dutch plant of Cincinnati Milacron, Dirk was hired by MDSI–France to serve customers in Belgium, the Netherlands, and Luxembourg. He still remembers arriving at the building on Plymouth Road for his first day of training: "There it was, that impressive-looking MDSI headquarters building, the deep blue sky behind it, the trees around it covered with snow, and the three tall poles with flags in front. That was certainly not how offices looked in Europe. I was really proud that, starting that day, I was a part of this organization."

Chet Fleszar saw the MDSI headquarters as "a true symbol of our success as a company." He particularly remembers watching Buildings II and III rise from the ground while he worked at his desk in Building I. Seeing that vision become reality felt to Chet like a rare and special experience "because it showed what hard effort can accomplish."

The leaders of MDSI were most proud of the fact that the nearly $30 million cost of the entire complex was paid in cash from operating profits only. There was no need for financing; MDSI had the money. When the new headquarters were completed, Ken told a reporter that it represented "a tribute to the efforts and dedication of MDSI's employees." Everyone had made this new home possible.

• • •

That included Seth's brother, David Powsner (#522), who worked at MDSI in the summer of 1978, after his first year at MIT. He was there as MDSI began its rapid growth in employees, averaging a new hire every working day for the next four years. It was David who helped the company figure out where those employees would be needed. That summer, Chuck said to him, "David, I'd like you to write a program to make a map of the state boundaries of the United States based on latitude–longitude coordinates."

David went to work programming the computer to make the map. When the map was complete, Chuck said, "Okay, now we want to make a table of ZIP Codes—the first three digits—with the latitude–longitude center of those ZIP Codes." When David got that going, Chuck pointed out, "For every one of our customers, we have a ZIP Code. So now we can see on the map where our customers are concentrated, which will tell us where we need to recruit new application engineers to service them." Chuck's team was ultimately able to total sales dollars by ZIP Code–center and draw circles on the map, each of which had an area proportional to sales dollars. With all these data to drive decision-making, MDSI knew where to establish regional offices; after a decade in business, the company had offices in thirty-two cities in the United States.

With so many employees out in the field, Mike Long developed an org chart of district managers and then regional managers. He eventually oversaw the work of more than 250 people in the field, but it never felt overwhelming. "I loved the hell out of that job," says Mike. He is proud to remember that, in 1977, MDSI achieved the milestone of having 10,000 NC machines under contract using COMPACT II.

By the fall of 1980, when MDSI was the cover story of *Ann Arbor Scene Magazine*, the company had more than 600 employees. Fully 76 percent of those employees had college degrees.[31]

• • •

HOT TECH COLD STEEL

One of the MDSI employees who didn't have a college degree was Teresa Killeen (#659). She was in her mid-twenties when she joined the company in 1977 as an assistant to special projects in the R&D group. Teresa took a course in blueprint reading for machine tools, did a lot of data entry, helped prepare customer manuals, and even traveled to a customer site in New Hampshire. When she became Don Redding's secretary, the tasks were more clerical, but she was also learning how to use the newest technology—the company's first fax machine (each page took seven minutes to process), an IBM personal computer, and a dot-matrix printer.

"I received my first email while working at MDSI," Teresa remembers. She was typing a memo on her PC when a message popped up on her screen and surprised her. "It said, 'Thanks for the onions.' I had grown onions in my garden at home, and I brought some into the office to give away." She learned that this message was "electronic mail," and she chuckles now at her memory of the first time she heard someone at MDSI shorten the name to "email": "I wondered why they couldn't just say 'electronic mail.'"

Teresa remembers MDSI as a "dynamic" work environment where people wanted to do the best job they could. The use of "peer review," says Teresa, "made people sharper at their jobs. We were encouraged to be the best we could be." Looking back on it, Teresa realizes that MDSI was exposing her to the concept of "best practices" before that was a trend. "It was a cultural expectation: we'd do the research to figure out what was the best way of doing something, and then we'd try to improve on it."

Being a female employee at MDSI put her in the minority, and Teresa agreed with the few female programmers who complained there was no excuse for so few women in R&D. But in the late 1970s, women were still scarce in computing classes, in machine shops, and in upper management. Though MDSI was a male-dominated corporation, Teresa says she never experienced inappropriate jokes or harassment. "That was probably because of Chuck's strong moral character," Teresa says today. "Everyone wanted to please Chuck, and he wouldn't respect people who conducted themselves that way."

Teresa remembers Chuck as "relentlessly cheerful" and "very accessible." She says he always came to new-employee orientations to give an enthusiastic pep talk. And she noticed that, even as MDSI's success must have meant financial success for Chuck, he still brought lunch from home every day. It impressed Teresa that the company's founder "could buck the insignia of success."

Though there were plenty of opportunities to advance in the growing company, Teresa says she didn't get to be a part of that, not because of her gender, but because she hadn't been to college. When someone else got a promotion she'd applied for—and knew she could do—Don Redding gave her a piece of advice that changed her trajectory: "You can spend your life being frustrated, or you can get your education." The seed was planted. Before long, Teresa was making new plans for her life.

■   ■   ■

In 1975, MDSI's board began to explore taking the company public. "Without my knowledge," says Ken, "they were talking to some investment banking people who knew how to do an IPO." When the company directors presented their plan to Ken, he was resistant to the idea—the work involved, the risk, and the loss of control of corporate decisions. But the investors were getting anxious to see a return. And Ken didn't have the authority to refuse.

"It required total immersion for weeks and weeks and weeks," he remembers. Working with Ehrmann and others, he developed reams of disclosure documents for the bankers. Ken was in New York City the day of the initial public offering in February 1976 when 787,500 shares were offered at $7.00 per share as over-the-counter stock (symbol: MDSY). For the rest of the time that Ken was president of the company, the price never fell below that initial value. In fact, for years, it just kept going up and up.

Investors began receiving dividends in the first quarter of 1978, and shareholders' equity went from $6.5 million in 1976 to $16.2 million in 1979. On September 18, 1980, the price was $59.50 per share.[32] Though the company would go through a lot of changes in the years following the IPO, a 1985 article in Forbes Magazine listed the 100 most successful IPOs in the years 1975–1985, and MDSI came in at #8 on the list.

This announcement is neither an offer to sell nor a solicitation of an offer to buy any of these securities.
The offer is made only by the Prospectus.

February 12, 1976

MANUFACTURING DATA SYSTEMS
INCORPORATED

750,000 Shares

Common Stock

Price $7.00 per Share

Copies of the Prospectus may be obtained from the undersigned only in states where the undersigned may legally offer these securities in compliance with the securities laws thereof.

C. E. UNTERBERG, TOWBIN CO.	HAMBRECHT & QUIST	PRESCOTT, BALL & TURBEN
BLYTH EASTMAN DILLON & CO. Incorporated		DREXEL BURNHAM & CO. Incorporated
E. F. HUTTON & COMPANY INC.	LEHMAN BROTHERS Incorporated	LOEB, RHOADES & CO.
PAINE, WEBBER, JACKSON & CURTIS Incorporated		REYNOLDS SECURITIES INC.
WHITE, WELD & CO. Incorporated		FIRST OF MICHIGAN CORPORATION
WATLING, LERCHEN & CO. Incorporated		MANLEY, BENNETT, McDONALD & CO.
WM. C. RONEY & CO.		SMITH, HAGUE & CO Incorporated

The Ann Arbor News, Thursday, February 12, 1976 39

Announcement of MDSI's initial public offering.

Despite this success, Ken tried not to get too excited about this metric when immersed in the day-to-day leadership of the company. "If I looked at the stock once a month, that was a lot," he remembers.

On the other hand, many employees found themselves looking at the stock price regularly, something they'd never done before. Dirk Van Krimpen, the AE in Europe, remembers a time when employees were offered a limited amount of stock options or a one-time cash payment. Dirk was already happy with the higher salary that his job at MDSI provided, and he didn't know anything about stocks. But he sought advice from the head of MDSI–France, John Adams, who told him, "Making money is not that difficult. But keeping money, that's really tough. I advise you to take the stock options."

Dirk followed the advice. He reflects, "I don't work any harder or try to be more clever because someone pays extra money for it. For me, the only criterion is whether it is fun to do. If so, I do everything I can to make my task a success. However, during the following weeks I started reading the *Wall Street Journal* to check the MDSI stock prices. And they continued to rise. It made us certainly aware of the impact our own work had on these prices. Most of us now slowly started to think of how our work could be made more profitable. And although not the most favorable topic, we also started to think of how we could reduce some of our cost. It was in our own favor to help the stock price grow."

Teresa Killeen also knew nothing about stocks when the company offered employees a Christmas bonus choice: $100 cash, or $100 of MDSY stock, or an option on 100 shares. Teresa asked Bruce Nourse for advice, and he told her, "Unless you absolutely have to have $100 right now, take the option." She did.

When she was later accepted at the University of Michigan—to get that college degree that would open more doors for her—she cashed in her 100 shares for something like $13,000. "I was working as a secretary," Teresa says. "I was living in an efficiency apartment. I had no financial plan for college. And now I had money to pay for my first year." Teresa left MDSI to study at U-M, graduated at age thirty, went on to get a law degree, and became a judicial attorney for the Washtenaw County Trial Court. Teresa credits her MDSI years for helping her develop good work habits, comfort with ever-changing technology, and a belief in herself.

■ ■ ■

Every employee got stock options, from the team leaders to the secretaries to the custodial staff. "I wanted them to feel part of the company," says Ken. "I thought people ought to have something of their own to work for." It also helped hold people to higher expectations and standards. "If a new employee came on board and didn't perform, other employees would tell him, 'Hey, Joe, get on the ball or leave. You're messing with my company.'"

It's not an exaggeration to say that most MDSI employees felt this was the best job they ever had. One afternoon, Ken was walking down a street in London with one of his application engineers, on their way to visit a customer, when Ken asked, "Well, you've been with us for a while now. How do you feel about your decision?"

The AE stopped walking, turned to look at Ken, and said, "A man only makes half a dozen major decisions in his life. He can probably count them on one hand. This is the biggest and best decision I've ever made."

Ken felt the same way. "MDSI was wonderful," he remembers. "It was like a fairy tale. The team spirit was what made it work. Success breeds success, and those people were so successful, they went well beyond what they thought they could do."

They also felt like a team when Ken gave them each a voice in solving company challenges. He says he still hears from former employees who remember the sense of being heard and valued in staff meetings. "I'll get an email from someone who was a junior engineer at MDSI, and they will say, 'I remember when you asked for my opinion about what to do. That meant so much to me. It's the same process I use today with my employees.'"

MDSI also invested in professional management training a couple of times a year. These one- or two-day courses helped the company's managers improve their ability to create a positive workplace where everyone could do his or her best. Ken, too, learned from these trainings, as management practices continued to evolve. He still remembers one seminar about the different personalities on a team— the achiever, the pleaser, the avoider, and so forth—and how to adapt management strategies to each psychological type.

■　■　■

Another indicator of MDSI's success was a decision in 1976 by the American National Standards Institute (ANSI) to name COMPACT II a national standard language for programming NC machines. Ken believes it was the first time a proprietary computer language (not developed in coordination with a federal research program) was made

ANN ARBOR
SCENE MAGAZINE

$1.25   FALL, 1980

MDSI's   Kenneth R. Stephanz   ●   Travel Feature: Haiti
The 1980-81 Professional Theatre Season   ●   The U. of M's Ray L. Fisher

Ken outside MDSI's headquarters, fall 1981. Photo: Lanny Lincoln Robbins.

a national standard. He says, "They did that because we had such widespread use compared to any other language."

According to Ken's calculations, there were about fifty-five competing NC programming systems in the United States, and another forty in Europe, at the time MDSI started. Ten years later, MDSI owned about 70 percent of the total market, more than all of its competition combined.

In Ken's view, another marker of success was the corporate jet. For frequent domestic business trips, the company had already purchased a twin-engine aircraft and then a Mitsubishi MU-2 turboprop. But in 1979, Ken said to the board of directors, "We'd like to get a jet."

It was an expensive request and maybe seemed a bit self-indulgent, since they all knew that Ken loved to pilot the company planes and no doubt sought the thrill of piloting his own small jet. So they used the request as an incentive, saying, "When MDSI achieves pre-tax earnings of $10 million, we'll authorize the purchase of a jet."

"I'm sure they were thinking we'd hit that goal in four or five years," Ken remembers. "One year later, we were there. I said, 'Give me my jet.'"

A company profile in a local magazine highlighted this purchase: "As a necessity of being a fast-track, computer-age company, MDSI owns a Cessna Citation jet aircraft, which Ken pilots in the same precise manner with which he runs MDSI."[33] The necessity of a corporate jet might have been up for debate, but it did get Ken to meetings around the country more quickly (he didn't use it to fly internationally). And the experience of flying it was an exhilarating perk. "It was much easier to fly than the turboprop," Ken reflects.

MDSI kept a pilot on staff, Albert Rohmann (#383), who flew other MDSI executives to meetings, and who served as co-pilot when Ken was the pilot. "He did nothing other than twiddle his thumbs while I was flying," Ken remembers. "I used to kid at our company meetings that, following on the bestselling book *God Is My Co-Pilot*, our pilot, Al, had written a new book: *Co-Pilot to God*." It was a bad joke, but one that Ken's colleagues remember to this day.

Al Rohmann was the pilot one winter night when Mike Long, Dick Stitt, and Rex Wolf needed to get to Massachusetts to meet with representatives from DEC. Al flew them in the MU-2 to Hanscom

Field, outside Boston. As Mike recalls, "It was late when we arrived and checked into a small hotel. The only place to get a late-night meal was a Howard Johnson's." As they ate, a light snow was falling, and by morning, it was snowing heavily. They made it to the meeting at DEC, but many of the people they were supposed to meet had been unable to get there in the snow. "About thirty minutes into the meeting," Mike remembers, "the PA system announced that a major snowstorm was underway and DEC was shutting down. We contacted Al and told him to get ready to leave. He told us we weren't going anywhere. The MU-2 was snowbound." The men checked back into the hotel and were stuck there for two days. "We played a lot of poker while waiting for the weather to clear." When it finally did, Al still couldn't take off because the runway hadn't been plowed. The MDSI men were at the end of their patience. Says Mike, "We left Al with the plane and flew home commercial from Boston's Logan Airport."

■   ■   ■

MDSI's board continued to be the same small group (Morgenthaler, Pavey, and Simon, plus Ken), and they must have been as amazed as anyone by the unabated rise in revenues throughout the 1970s. When they started this venture in 1969, they were optimistic about its prospects. But they could hardly have expected the company would meet or exceed annual financial projections for ten years running.

And that wasn't because the projections were conservative. Ken remembers, "Every year, when I got through tweaking the next year's budget, I'd think, *Hmmm, I'm not sure we can deliver on this.* But then we would."

By the end of MDSI's first decade, customers numbered more than 3,400, including multiple Fortune 500 companies. The program to teach customers to use COMPACT II was training nearly 3,000 people a year. And gross revenue for 1979 was $42,546,000, with after-tax net earnings of $4,295,000.[34]

"It was magical, truly magical," says Ken. "All previous steps in my career paled substantially in comparison to the MDSI experience.

Not just the phenomenal growth, but the fun, too. It was the best entertainment ever."

But as the seventies came to a close, there were hints around the edges that the party was starting to wane. And then Chuck Hutchins called it quits.

Part IV
The Wind Up

17

The Cash Out

Perhaps a clash between MDSI's two founding executives was inevitable. Though Ken Stephanz and Chuck Hutchins were both engineers, they clearly had different ways of working and managing employees. And they saw the company from two different vantage points.

As a salesman and corporate leader, Ken was the guy who learned early in his career that "the answer is yes." If customers or investors asked for something he wasn't already doing, it was always better to agree to do it and then figure out how.

But Chuck, in his role as vice president of research and development, was the guy who had to find those solutions, and he saw innovation as incremental. Each step in his career was built on the step before it. Chuck was as curious as ever, and excited about advances in the field, but he wasn't inclined to make promises he couldn't keep.

Celebrating MDSI's tenth anniversary, February 1, 1979. From left: Chuck Hutchins, Ken Stephanz, George Shingler (#266), Fred Stevenson (#267), Chris Neylon (#260), Bruce Nourse, and Urbanes Van Bemden.

By 1978, with an ever-expanding team of computer engineers working at the cutting edge of software design, Chuck was becoming wary of customer requests that threatened to outstrip his team's capacity and capabilities. "To write software, you must know about the subject you are programming," says Chuck. "Ken started telling customers that we could basically do anything. I warned Ken. I told him that when a rocket goes up and doesn't reach escape velocity, it comes back down." Chuck told Ken he wasn't inclined to stick around if that was where things were headed. "But," Chuck remembers, "he talked me into staying."

■　■　■

It wasn't that Chuck was clinging to outdated strategies. In fact, one of the last things Chuck worked on was the project to write an early version of 3-D CAD software. He remembers, "Bruce and I went to Lockheed in California to learn about a program called CATIA." CATIA (Computer-Aided Three-dimensional Interactive Application) was developed in 1977 by a French aircraft manufacturer for in-house use, and Lockheed had adopted it. "Bruce and I spent three days out there, reading their manuals and so on, and we came back, and I said, 'Bruce, it appears to me that the only difference between what CATIA has and what we have is they're on a 32-bit computer, and we're on a 24, but I don't see that we'd gain anything by that.'"

What Chuck could see, at that point, was the future of computer-aided design: "CAD wasn't going to be lines and circles anymore; it was going to be 3-D solids. And that's when we created DESIGN, the Design Engineering System Including Graphic Notation."

That graphic notation allowed programming for planar, cylindrical, conical, and spherical objects. Says Chuck, "With the help of Bob Johnson and others, we got DESIGN to the point where we could write a program to design a hydraulic cylinder. To do that, the software required three parameters—the pressure, the force, and the stroke—and out would come the design for a complete hydraulic cylinder."

This progress in programming 3-D CAD caught the attention

of at least one stock brokerage firm. In a letter to their investors recommending emerging growth stocks in high technology, the Boston firm of Adams, Harkness & Hill, Inc. noted that MDSY (MDSI's stock symbol) looked like a good bet: "The CAD/CAM market is a huge one. . . . On this front, the company has development projects in most significant CAM areas and is beginning to come up with quite a few products. . . There is even MDSY's own 'pot of gold at the end of the rainbow'—a project which would provide real 3-D CAD instead of the 'wire-frame' approximations of all current systems."[35]

As MDSI's board examined the CAD potential for the company, they considered merging MDSI with a leader in the CAD field, Applicon. Founded in Massachusetts in 1969 (the same year MDSI was founded) by a group of MIT computer scientists, Applicon produced computer-aided design software that ran on the DEC PDP-11 minicomputer. One patented Applicon innovation was the ability to input commands using "drawn character recognition" with a stylus and tablet that connected to the computer.

David Morgenthaler reached out to his friends at J.H. Whitney & Co. (the firm that had originally inspired Dave to become a venture capitalist) to facilitate a conversation with Applicon. But each company had a strong CEO who wanted to command a post-merger company, and talks fell apart.

Chuck still believes that MDSI's DESIGN software was ahead of where Applicon was at the time. "That's one of the tragedies of my story," says Chuck. "I really thought we were onto something. We should have been the world's next CAD company."

■ ■ ■

DESIGN was the kind of innovation that showed Chuck was thinking long-term about meeting customer demands. But more immediate changes were afoot at MDSI. In April 1979, Ken promoted Chuck to executive vice president. Then he introduced Chuck to the new vice president of R&D, Barry Borgerson (#1025). The changes weren't a complete surprise, since Chuck had asked for this kind of restructuring two years earlier. But for Chuck, the timing and the way it happened seemed intended to diminish his influence.

Shortly thereafter, the executives of the company held their annual spring retreat, this time at a resort on Lake Erie. Ken chose this more affordable location when he decided to expand the attendance to include all the middle managers as well. Chuck remembers, "So many good ideas came out of those meetings, enough to easily make a five-year plan! Cai Raber did a superb job of recording all the ideas."

When the Lake Erie retreat participants returned to the office, Chuck told Ken that his key takeaway from the retreat was that MDSI needed to develop a new business plan. He stressed his belief that the plan should be ready by September 1, 1979.

Such a business plan had a lot to consider. In the rapidly changing world of computing, MDSI's future faced some clear challenges. The success of the Action Central minicomputers in serving customer needs with a one-time purchase meant the loss of continuous revenue from timesharing. And as stand-alone computers became smaller and more affordable, everyone could see that the era of computer timesharing was fading.

In addition, advancements in the controllers attached to each machine tool were also likely to compete with MDSI's business model. "In the early years," Bob Pavey explains, "the tapes produced with COMPACT II were fed into dumb controllers. But over time, the controllers got smarter. Customers were starting to make changes to their programs directly on the controllers. It was clear that this trend was going to continue." Cai Raber's team found that these smarter controllers reduced the need for MDSI to write machine tool links for customers. As Pavey remembers it, the MDSI board members saw these trends coming and knew that "we didn't want to be in the minicomputer business or in the controller business."

Mike Long remembers, "It was easy enough to see that, over time, the revenue base was going to be eroded by continual minicomputer sales. That led to quite a bit of tension and, basically, a falling out between Ken and me. I felt that without a business plan that incorporated moving forward with new products and services and more enthusiastic support for stand-alone systems among customers, the future of the company was going to get pretty cloudy." Long resigned from MDSI that spring of 1979.

By the end of June, Chuck saw no sign that a new business plan was even being considered. For him, that was the last straw. "On June 29, 1979, I shook Ken's hand and quit."

■ ■ ■

Ken remembers these resignations slightly differently, and MDSI's continued success in the short term bolstered his sense that the company was still on the right track. When MDSI was profiled in a lengthy cover article in the Fall 1980 issue of *Ann Arbor Scene Magazine*, there was no mention of Chuck's or Mike's departure or of any future instability in the company. Although the magazine—a locally produced business promotion and advertisement circular—was not meant to offer hard-hitting journalism, it *was* widely read in the community. And since MDSI was one of Ann Arbor's largest employers, local leaders were counting on the rosy projections for MDSI.

The article mentioned three new products that MDSI introduced at the September 1980 Machine Tool Show in Chicago: "a programming system to assist manual programmers; a job estimating, costing, and general accounting system for small job shops; and a computer-aided drafting system for draftsmen and engineers."

Ken told the magazine's reporter, "We're broadening our product base by building upon the leadership position that we've established in the NC field. These new products exemplify our interest in extending MDSI's manufacturing knowledge and computer-assist expertise to help increase productivity in other areas of manufacturing operations."

The article continued, "To better prepare the company for the opportunities which will ensure sustained profitable growth in the 1980s and beyond, the company recently took a number of steps. Organizational restructuring was done to facilitate expansion and diversification for the company's product line. Strengthening of the management team was also accomplished along with the implementation of ongoing management development programs. The number of field salesmen and application engineers has also been increased dramatically."

For Bruce Nourse, this description was business jargon for painful changes in upper management. With Chuck and Mike gone, new VPs from outside the company ushered in a new era at MDSI. Says Bruce, "A whole bunch of people came in who had not been there from the beginning. They were now the top management, but they didn't know the history of the company, they didn't know the culture of the company."

Cai Raber, head of the link-writing group and who had joined the company in 1969, also saw that things were changing. "When Chuck left, as far as I was concerned, that was the handwriting on the wall. MDSI had a good history, but I had a feeling it was time for me to go." Cai left Ann Arbor for a job in Hartford, Connecticut, driving across the country with his wife and infant twins, pulling a trailer behind his car. He was sorry to leave, but he wasn't bitter about it. Remembering that time now, at age eighty-two, he shrugs: "All good things come to an end."

■　■　■

Even as Ken was talking to that magazine, he was also in talks about the possible sale of MDSI to the international oil field services giant Schlumberger. MDSI's directors had been exploring this idea even before the company went public in 1976. At that time, the investment bankers involved with the IPO had talked to the chairman of Schlumberger, a Frenchman named Jean Riboud, about a possible acquisition.

They revisited the idea after the Applicon deal failed to materialize. "Schlumberger was aware of us," Ken remembers, "and Schlumberger acquired a lot of companies in that period of time."

Schlumberger's reasons for wanting to acquire MDSI are not fully known, but a company report shows that Schlumberger wanted to tap into the burgeoning CAD/CAM market in order to stay competitive as computers transformed industry: "Computers are finding their way into engineering departments and onto factory floors . . . Many observers liken these advances to a new industrial revolution. . . The ultimate goal of this revolution could be a fully automated factory, and helping to bring this about are companies which produce CAD/CAM (Computer Aided Design/Computer Aided Manufacturing) systems. . . . A huge market remains to be tapped in CAD/CAM. . . . It's estimated that CAD saturation amounts to less than 5% of the potential market. One day engineers will use CAD/CAM systems as widely as they now use calculators."[36]

With this kind of potential in mind, Riboud came to Ann Arbor and spent a whole day with Ken. "He was a wonderful guy," Ken remembers. "He saw what we were all about and what I was all about." Ken could see that acquisition by Schlumberger would greatly increase MDSI's international recognition and status.

But Ken had mixed feelings about where things were headed. "MDSI was my baby," he says. "I didn't want to give up anything of it." The problem was, MDSI wasn't Ken's company. It was the shareholders' company, the investors' company. Ken's personal feelings didn't enter into the deal. An article in the September 19, 1980, edition of the *Wall Street Journal* announced the plan for the acquisition, with an expected price of $189 million. (MDSI's 1980 gross revenue was $56 million, with after-tax net income of $5.4 million.[37])

Ken accompanied Dave Morgenthaler to New York City to negotiate with Schlumberger. The French company's North American headquarters had been in Houston since the 1930s, but in the 1970s, its top executives relocated to New York. Jean Riboud was there to meet Ken and Dave, as were lawyers for both sides.

By the end of the day, it looked like they had a final agreement on a price, and Dave and Ken returned to their hotel. The deal was a good one, but Ken wasn't happy; he felt that if he had to give up his "baby," he wanted to push for more money. "I had an obligation to our shareholders," he remembers.

At the hotel, he turned to his chairman and said, "Dave, I'd like to have a sit-down meeting with Jean Riboud, just the two of us." Dave had just succeeded in turning his modest initial investment into the best deal of his venture capital career to date, but after more than a decade working with Ken, he also trusted and respected his CEO. Dave agreed, and the next day, Ken got his one-on-one meeting with Riboud, resulting in a substantial increase in the sale price. When the shareholders approved the deal on January 21, 1981, Schlumberger paid $210 million for MDSI (about $650 million in 2020 dollars).

For Dave Morgenthaler it was a spectacular result. His original investment of $200,000 turned into a personal payout of $20 million. The success of MDSI gave Morgenthaler immense credibility and visibility in the venture capital world. He co-founded and served as chairman of the National Venture Capital Association, and through that role, helped to persuade Congress to lower the capital gains tax rate from 49 to 28 percent and to change the law to allow pension funds to invest in venture capital firms.

Bob Pavey, his partner at Morgenthaler Associates, made his first million on the deal. He credits Dave with a key role in growing the venture industry and private equity from $100 million in 1978 to $200 billion in 2000. When Dave died in 2016 after a long lifetime of successful investments, the obituary written by a reporter at the *New York Times* featured Manufacturing Data Systems, Inc. as the investment that "netted him his fortune."[38]

■　■　■

Ken called an all-company meeting to announce the sale of MDSI to Schlumberger. As he remembers it, the news was met with enthusiasm, because the employees had stock options. "Everyone was going to make money on the deal," says Ken. "I made several people millionaires."

Jean Riboud, chairman of Schlumberger, touted the acquisition of MDSI in his February 1981 letter to shareholders (published in Schlumberger's 1980 Annual Report). Noting that this was his company's entry into CAM, he praised MDSI for having "built a very special position, both marketwise and in software capability with the large number of numerically controlled machine tool users." According to the report, MDSI had 3,850 customers and 18,500 NC machine tools

Ken (center) with Schlumberger leaders, signing agreement for Schlumberger to acquire MDSI, 1981.

under contract. Said Riboud, "This is the beginning of a fascinating new adventure."

Not everyone was thrilled. The Boston brokerage firm of Adams, Harkness & Hill, Inc., which had been consistently recommending MDSY stock to its investors, sent out this announcement in January 1981: "We don't begrudge the money holders will make, nor regret the fact that we recommended the stock on two different occasions in the last year. All the same, we spent all Friday moping around the office after learning about the agreement that Schlumberger would buy out this company. 'What a shame!' we thought. In our view, MDSY was *the* productivity play. We had been looking forward to a long and profitable association with this company in the next few years. We do not like losing such companies!"[39]

At the time of the Schlumberger deal, MDSI had almost 700 employees worldwide (with 500 in Ann Arbor).[40] Ken believed they had no reason to fear for the future of their jobs. "Keeping the employees was an obvious thing for them to do," he says, "because Schlumberger did not have expertise in what MDSI was doing."

As for his own job, Ken was on his way out. The acquisition deal included naming Ken to a senior vice president position in Schlumberger's New York City offices. That promotion offset some of his disappointment about selling off the company he'd built. So did the millions he earned from his own MDSI stock options.

Chuck, too, made out well with the sale of his shares, though he was long gone from the company by then, and he was dubious about the acquisition. "When the directors sold the company to Schlumberger," he says, "they knew it was in trouble. And it started going downhill right away."

18

The Fall Out

MDSI's long, slow demise wasn't immediately obvious to its employees. In fact, new hires continued to join the MDSI team in 1981 and 1982, and business continued as usual for much of that time.

However, when Schlumberger acquired Applicon the year after acquiring MDSI, they shut down development of DESIGN—the 3-D CAD software project. Whereas MDSI's R&D team had spent much of the 1970s advancing technology for both computer-aided manufacturing and computer-aided design, Schlumberger considered Applicon the CAD company and decided that MDSI's focus should be narrowed back to its original CAM expertise.

When the Applicon deal was finalized in January 1982, MDSI reportedly employed a total of 850 people, with 550 working in Ann Arbor, while Applicon had 1,000 employees in its Burlington, Massachusetts, location.[41] Schlumberger created a new corporate unit called "Computer Aided Systems" to oversee the work of all these employees, headed by former Applicon president Donald Feddersen.

In an article about the pairing of the two companies in the *Ann Arbor News*, Schlumberger spokesman Sid McCormick said that the two firms "would not lose their individual names or identities or face plant closings or loss of business due to the consolidation."

The *News* reporter reached Ken in California for a comment: "The synergy with the two companies will be outstanding," he said. "This will put us in the forefront of the CAD/CAM field, and that's a basic strategy that I'd been promoting for several years. Once we joined Schlumberger, we had the resources to do it."

But those resources quickly began to diminish. According to Schlumberger's 1983 annual report, "shipments" from MDSI to customers were down 55% in 1983, and revenue was down 16 percent.

Bruce Nourse remembers feeling "aggravated" by the new leadership at the company he'd helped to found. "Schlumberger came in and put a bunch of top managers in place who didn't understand the culture of the company at all. The year they bought it, that was the last profitable year the company had." Bruce's warnings about problems with software projects went unheeded, and his new supervisors seemed more interested in maintaining protocol than hearing feedback. He remembers, "I was there when MDSI was making $60 million dollars a year, and I was there when it was losing a million dollars a month." By 1983, he was "done with it." He found another job—at half his MDSI salary—and resigned.

As Bruce was preparing to depart, Betty Ruddy created a scrapbook for him (as she had also done for Ken) of MDSI brochures, newspaper clippings, trade articles, and photos that documented his long tenure at the start-up he helped to found. Betty would retire from Schlumberger in 1986 at age sixty-one.

■ ■ ■

Meanwhile, the talented team of computer engineers at MDSI continued to work on interesting projects as Schlumberger sought to diversify the offerings of its new holdings.

After six years of heading the MIS department, Ron Peterson transferred to a new product development team headed by Peter Cullinan (#990) that was working on a product called Comshop. The intended product was a computer system that would perform any and all business functions required to run a machine shop, including accounting and job cost estimating. But after months of effort, says Ron, "funding for the project ran out before we could deliver a solution to the customers."

Ron was then sent to manage another product development team, working on CAPP (Computer Assisted Process Planning) using CODE (the graphics-based database system that MDSI had acquired in 1974). According to Ron Peterson, the new CAPP product would be "completely portable to most foreign countries by having all textual data that were part of the program driven by a library of messages which could be edited for any language."

The CODE / CAPP project was the first assignment for Richard Sheridan (#1300) when he arrived as an intern in May 1980. It was the summer before his last semester as an undergraduate computer science major at U-M. During intern training, Rich remembers that U-M adjunct professor John Sayler and MDSI employee John Samford "taught us the 'right way' to build code using Pascal, my first exposure to the language. They taught us how to keep the modules small, well structured and well documented, and interestingly, how to do test-driven development: write the tests, write the code, run the tests, over and over again. Our industry wouldn't come close to adopting this as a standard for another twenty to thirty years. I was in heaven. This was paradise for a computer science major."

But after that training, Rich reported to Vince Bobrowicz (#306) of the CODE / CAPP team. Says Rich, "Vince pulled me into his office, closed the door, and told me that I should promptly forget everything I just learned in that class. The real world came crashing down around me very quickly." The project required Rich to work in a language called NFO, which stood for "Not Fortran." "It was structured Fortran (goto-less) that was then run through the NFO preprocessor, turned into regular Fortran (with goto's every third line), and then compiled. What an abomination. It was almost impossible to debug, as all debugging work had to be done with the output of the preprocessor." In addition, Rich saw some examples of management styles he would not want to emulate if he was ever in a leadership role. But he soon mastered the NFO process and began to make some kind of organization out of the mess. "I actually started to have fun," he remembers.

He marveled at another young and "massively fast" programmer named Tom Knoll (#643), who would eventually create Adobe Photoshop. And he met a "calm and talented" programmer named Gloria Page (#1873), a recent computer science graduate from Michigan State University who sometimes brought her young son, Lawrence, to visit the office. That Michigan native would grow up to be Larry Page, co-founder of Google.

Rich and his coworkers played the card game hearts during lunch breaks in the cafeteria, and some Fridays they would leave for lunch, drink all afternoon, and never go back to the office. "It didn't seem unusual given the crazy hours we worked otherwise."

Rich continued to work at MDSI while pursuing his master's degree at U-M, and as he prepared to graduate in the spring of 1982, he had job offers from Bell Labs in Naperville, Hewlett-Packard in Cupertino, Data General in Westboro, Standard Oil in California, and MDSI (which, by then, was a Schlumberger company). MDSI had invited him to join a major effort to rewrite CODE and CAPP in Pascal. "It would be an exciting new project, a chance for a new grad to lead an effort to build something great and lasting. I chose MDSI." Looking back on that time, Rich says, "There were so many wonderful choices, but . . . I really wanted to stay in Ann Arbor. It seemed like such a great town to live in and raise a family in. I have never regretted that decision."

But MDSI would soon give him reason to regret his choice of employer.

■　■　■

In spring of 1982, Ron Peterson and his CODE / CAPP team were preparing for their first trade show with the new product when Schlumberger suddenly cut their funding, halting all related work.

Rich Sheridan had taken his wife on a six-week trip to Europe just before starting his full-time job at MDSI when the news broke. "As soon as I arrived back, the layoffs began." Rich wasn't laid off, but he remembers, "I kinda freaked out, as I had just turned down all these other great job offers."

MDSI employees now remember that day in June 1982 as "Black Friday" when approximately 250 people were let go. Ron Peterson was one of them, but he remembers that "MDSI management did a remarkable thing in providing professional outplacement services with consultation, secretarial staff help, postage, and phone access to help us find another position."

Rich Sheridan stayed on for a while, and the company found other programming tasks for him to do. "In early 1984, I was assigned to work on a new version of NC Graphics, which basically generated COMPACT II using graphics for both parts and tool paths." Apple had just released the first Macintosh computer, and Rich tried (unsuccessfully) to get NC Graphics working on a Macintosh.

He believed he would be able to do a better job with these efforts if he could talk to customers, so he asked his supervisor how that could be arranged. Rich remembers, "He told me if I didn't stop bugging him about talking to customers, he would 'stick me' in customer service. I said that would be great! I would get to talk to customers all day and hear of their problems and then be able to effectively design a system that didn't have those problems." Rich's supervisor didn't see it that way. He told Rich, "You don't understand. If I stick you in customer service, you will never come out again. You will never be able to write code again at MDSI."

Rich remembers, "I found a new job at Winterhalter, Inc., in less than thirty days." Al Kortesoja tried to talk Sheridan out of leaving, but Rich could see the writing on the wall. "MDSI was doomed. It would never be the same again." Rich Sheridan went on to found Menlo Innovations, an Ann Arbor software company known for its unusually happy, energetic, and creative teams. He has even written a popular book called *Joy, Inc: How We Built a Workplace People Love*, and he speaks all over the world on what he's learned about creating joyful workplaces.

■   ■   ■

Wayne Esch (#1816) was added to MDSI's management information systems team in April 1982, just weeks before Black Friday and Schlumberger's hiring freeze. (The last employee number, according to a spreadsheet Chuck created years later, was #1878.) As a hardware specialist whose most recent job was director of business systems for Domino's Pizza, Wayne was hired to develop MDSI's first in-house computer center. After twelve years of using timeshare computer services for their own computing needs, MDSI under Schlumberger was finally making the leap to its own servers.

Wayne remembers, "Dick Stitt provided an empty space in the basement of Building III with sufficient air conditioning and a raised floor for the in-house computer center." After installing DEC System-10,

Data General, and SUN servers, Wayne and his team had to migrate all the engineering software development from the timesharing services at ADP to MDSI's own computer system. "I had the further responsibility of developing network connections to offices in the U.S. and overseas," says Wayne. "I installed leased lines to U.S. offices and connections to the international offices using X25 and X75 Gateway networks, so that U.S. and international offices could access the new computer center in Ann Arbor."

As the years progressed, Wayne would eventually install and manage VAX, Auspex, NAS, Apple, and PC computer hardware in the MDSI offices, and implement internet, intranet, and email services company-wide. His operations staff (which numbered twenty-four people at its height) became what we would now recognize as a typical in-house IT department. They installed and maintained hardware, software, networks, workstations, servers, printers, network cabling, switches, bridges, routers, patch panels, and telephone PBXs. They also backed up and restored computers, designed standard cubicle layouts and moved cubicles, inventoried all equipment with MDSI serial numbers, sold surplus equipment, and reviewed security cameras.

Wayne remembers that the office culture still included entertaining staff events, like chili cook-offs, softball games, and racquetball tournaments. And the R&D guys continued to be a little unusual; Wayne knew a programmer who often slept in the stairwell.

He also remembers multiple managers from Schlumberger transferring in for a few years and then leaving. "I reported to sixteen different managers," says Wayne.

A perusal of Schlumberger's annual reports from the 1980s suggests that the multinational oil field service company never quite figured out what to do with its "Computer Aided Systems" division. Operations were "regrouped" in 1984, and in 1987, a few other companies were added to the division, including Sentry, Factron, and Benson. In 1987, the division was renamed "Schlumberger Technologies." Despite these efforts, the work of these holdings never seemed to be integrated into Schlumberger's central business.

• • •

Meanwhile, MDSI employees throughout the world were adjusting to the changing dynamics of a company on the downslope. While Ken Stephanz had been committed to working through wholly owned subsidiaries, Schlumberger encouraged the use of distributors. Soon enough, Urbanes "Van" Van Bemden—who had already worked for years with the MDSI offices in Europe and Japan—found himself working through distributors in South Korea, Hong Kong, Taiwan, Singapore, New Zealand, and Australia.

At that time, sales of MDSI products internationally were still fairly robust. Van remembers participating in an international sales contest in 1984, organized by MDSI's head of international operations, Steve Imredy. Says Van, "I had a potential customer in Christchurch by the name of Mace Engineering, the largest machine shop in New Zealand. I had been working on them for some time to buy a minicomputer system. Whenever I went to Australia to work with our distributor there, I would stop off in New Zealand to try to make this sale. As the sales contest was coming to the end, I telexed Ted Mace, the son of the owner, who was pretty much in charge of Mace Engineering, and suggested that if he was going to place an order with us any time soon, I would love to have him place it now. He did. The addition of his order allowed me to win the International Salesman of the Year contest and a $6,000 prize."

Not long after that, Van was ready to transition out of MDSI's international division. In the process of orienting his replacement, Gordon Creer (#1785), Van had to introduce him to the various distributors and customers. "Thus, he and I and our wives were off to the Far East," Van remembers. "While in New Zealand, we called on Mace Engineering, where I picked up another $25,000 order." During that meeting, the elder Mr. Mace asked what Van and his companions were doing that weekend and suggested the group visit Milford Sound—a stunning convergence of mountains and waterfalls and rugged coastline. As a board member with Air New Zealand, Mr. Mace arranged a fifty-passenger plane to take the four Michiganders to

Milford Sound. For Van, it was a memorable conclusion to more than a decade of traveling the globe.

G. Howard Barrett (#208), a resident of West Yorkshire, England, had been with MDSI since 1974, working mostly in the highly industrialized northern U.K. When the company was growing like wildfire, he found himself managing more and more salesmen and application engineers, a role that was new for him. "I was doing it with my own initiative," Howard remembers. "I said, 'I need some proper management training.'" About that point, Schlumberger had bought the company, and they offered to pay Howard to get an MBA.

He had completed two years of a three-year course when Steve Imredy, the head of MDSI's international division, asked Howard to take over MDSI–Sweden. "You'll get a year to either make it profitable or start to close it down," Steve told him.

Howard was hesitant. "I'm working on this MBA," he told Steve. "I can't leave now."

Steve said, "If you do this job properly, you won't need an MBA. If you don't do it, Schlumberger won't be paying for your schooling any longer anyway."

Howard decided he had no choice. By the time he started the new posting, Steve Imredy was gone from MDSI.

The tax on residents of Sweden was so unfavorable that Howard decided to commute to the job, rather than move his family there. He flew from England to Sweden forty-six times in a year. He became so familiar to the airline staff that they would offer him a first-class seat whenever one was available. One trip, he ended up sitting beside the two male members of the Swedish pop music group ABBA.

The Sweden office survived under Howard's leadership, enough so that he was asked to spend the next year at MDSI–Germany. This time, he moved his family with him. When Howard arrived to that position, he found that the head of MDSI–Germany had just left the company. Before the year was out, Howard, too, was keeping an eye on the exit. When he got a call from a former MDSI employee who was running a tech company out of San Diego, Howard jumped at the chance to direct that company's European operations. In 1986, Howard resigned from the former MDSI, which was now called Applicon after Schlumberger merged MDSI's operations into Applicon the previous year.[42]

. . .

On June 3, 1988, Schlumberger organized a party to honor Van's twentieth anniversary with the company. It was twenty years that month since Chuck Hutchins had hired Van for his small team at Comshare. Van was now the only founding employee still with the company. A small crowd gathered in the cafeteria to salute him, and company managers presented Van with a plaque and a Rolex watch.

The following year, Chet Fleszar retired from the company after twenty years in customer support. He was sixty-eight. "Schlumberger made it very difficult for me to retire," Chet remembers. "They asked me if I would stay on for another project, knowing full well that their projects normally went three years, and I would be in my seventies before it was done. I enjoyed the work, but by then my wife was retired." He gave his bosses three months' notice and retired in 1989.

Now ninety-six years old, Chet talked by phone about his memories of his MDSI years and reflected on his contributions: "I have always felt, even as a young person in military service, that we owe the country for the wonderful world we live in. Whatever we do, we have to look at the bigger picture. MDSI gave me that opportunity. I owe Ken and Chuck a debt of gratitude for taking a gamble on me and allowing me to meet my potential."

In the fall of 1992, there were still 300 employees working for the "Applicon" division of Schlumberger at the Plymouth Road offices of the former MDSI.[43] But the next year, Schlumberger sold what was left of its MDSI / Applicon combination to Gores Enterprise. That led to another round of layoffs, which included Van who officially was laid off on December 1, 1993.

Wayne Esch was still with the company when it became Gores Enterprise, and then when it became Unigraphics and then UGS Corp. He witnessed waves of layoffs over his twenty-one years, but somehow, as the guy in charge of the company's own IT needs, he was always kept on.

He was still there when the three buildings on Plymouth Road were sold to the University of Michigan in 1997 and what was left of the company moved into a smaller, rented office space.[44] Wayne was one of just thirty-two employees still on the payroll when, in 2002, he was finally laid off.

Bruce Nourse believes there are still a few users of COMPACT II to this day, running the software on old minicomputers in small metalworking shops. "It hasn't been updated in many, many decades," he says, "but it still works, it still does the job."

As for the beautiful corporate campus that MDSI built at 4251 Plymouth Road in Ann Arbor, it is now called Arbor Lakes and is the home of U-M's Information & Technology Services. Chuck likes to say that these are, "the only buildings on the University of Michigan campus that were paid for in cash thanks to a fateful handshake between two Michigan Engineering alumni—Bob Guise and Chuck Hutchins."

Photo: Dwight Burdette, CC BY 3.0 (https://creativecommons.org/licenses/by/3.0), via Wikimedia Commons.

HOT TECH COLD STEEL

EPILOGUE

In the fall of 2019, Chuck Hutchins was sitting in a coffee shop in downtown Ann Arbor when a self-driving robot passed by, outside. Chuck was in the midst of a meeting about this book, when he suddenly pointed out the window: "Look at that!" His compatriots and other patrons turned to see a sleek, silver, three-wheeled box, about the size of a mini-fridge, with no windows and, obviously, no human driver, tooling down the road, staying near the curb, on its determined way somewhere.

"That's one of the new, food delivery robots," someone said. A bicyclist was following behind, keeping an eye on the pilot project that connects restaurant takeout to offices and homes via a remotely monitored robotic vehicle.

When it was out of sight, Chuck returned to his meeting, but the boyish grin remained on his face. "The only benefit of being eighty-five," he said, "is that I've seen the whole show."

Chuck is still endlessly fascinated with computer technology, engineering marvels, and precision manufacturing. He loves to visit 3-D printing labs, and he is continually upgrading his home computer. Though he's not at all a braggart, he sees the effect that his career had on the future he's witnessed, especially in today's automated machine shops, and he's proud to tell the MDSI story and to talk up the work of his former colleagues to anyone who will listen.

With decades of hindsight, the wisdom in MDSI's business model is obvious. Gary Morgenthaler—son of David Morgenthaler and a current partner of Morgenthaler Ventures—sums it up this way:

> *Manufacturing Data Systems, Inc. (MDSI) pioneered today's primary computing paradigm; i.e., timesharing over dial-up modems in the 1970s was the equivalent to today's "cloud computing" on the internet. MDSI also pioneered today's primary software pricing model: "Software-as-a-Service" or "SaaS" with usage-based payments and not license fees. Finally, MDSI pioneered today's most successful business model, that is, the "platform" model*

on which third parties contribute software-value added on a core platform. MDSI's third parties provided their numerical control software via an API (application programming interface) using MDSI's COMPACT II programming language. The most successful software companies today are 1) cloud-based, 2) SaaS-oriented, and 3) using a platform business model. MDSI was nearly fifty years ahead of its time. These visionary choices account for much of its extraordinary success.

■ ■ ■

In 1981, the Numerical Control Society recognized the contribution of Chuck Hutchins to the development of computerized machine tools. That year, he was a co-winner of the Society's highest honor, the Joseph Marie Jacquard Memorial award (named after the eighteenth-century French weaver who is credited with developing the earliest programmable loom).

The other winners that year were two of Chuck's contemporaries: Richard A. Stitt (developer of the ACTION language and founder of NCCS, which became part of MDSI) and Harold Baeverstad (the Sundstrand employee who developed the original SPLIT language on which both ACTION and COMPACT were based). That award ceremony was an important marker in the chronology of the development of what we now call CNC (computer numerically controlled) machines.

In the years following that recognition, Chuck continued to explore the use of computers in manufacturing, as both a researcher and an investor. From his home in Florida, where he lived on a sailboat with his wife, Ann, for nine years, Chuck used his earnings from the sale of MDSI to support new business ideas.

One of those ideas came from the former head of Applicon, Don Feddersen, who introduced Chuck to a former Applicon employee, Samuel Geisberg. Geisberg, a Russian-born mathematician, had written the first "parametric, associative feature-based, solid modeling" CAD software. At Feddersen's invitation, Chuck consulted with

Geisberg for three days in Boston, and Chuck's enthusiastic support of Geisberg's ideas solidified Feddersen's plans to back Geisberg. This led to the 1985 founding of Parametric Technology with its main product, Pro/ENGINEER (Pro/E). Early customers included John Deere and Caterpillar, and Parametric (now PTC) eventually became a billion-dollar company. Says Chuck, "I was really saddened by the failure of MDSI to continue the work we had started on DESIGN, because that is where I thought the world was going. Parametric proved me right. We should have been that company."

■　■　■

Meanwhile, Bruce Nourse was working at an Ann Arbor company called Applied Intelligence Systems, Inc. (AISI). He had learned about the company from Jon Ehrmann, MDSI's former CFO, who had taken a job there. AISI built image-processing computers, which, Bruce points out, "have to be very, very fast, and the way that's done is with micro-coding." Bruce had gained some micro-coding experience at MDSI, so he was an invaluable addition to the AISI team. "All their applications used a common set of micro-code. Everybody used it, and I wrote it."

Bruce was AISI's principal designer and programmer for eight years. Then in 1991, he got a phone call from Chuck, who said, "I have another idea I want to work on. Do you want to work on it with me?"

Bruce remembers, "Chuck wanted me to write the program for a complete, numerically controlled controller, something that would read G and M code tapes and make the machine move." The two longtime friends and colleagues co-founded a new company together. At first, they called it Software Algorithms, Inc., but when Chuck discovered that Schlumberger had stopped using the name MDSI and had not continued to protect the name, he decided to take it back, counting on MDSI's previously stellar reputation to push the new venture forward. "MDSI2" lasted seven years with a focus on, as their promotional material said, "open-architecture CNC, motion control, and factory automation solutions."

OpenCNC®
SOFTWARE CNC

OpenCNC® is a production-proven, unbundled, software CNC control built on an open architecture that enables manufacturers to integrate off-the-shelf hardware and software technologies to achieve flexible and agile world-class manufacturing. Unlike proprietary CNC controls, OpenCNC requires no proprietary hardware or motion control cards. Combining a soft CNC and soft PLC in a single application, OpenCNC is well suited for new equipment as well as machine control replacements.

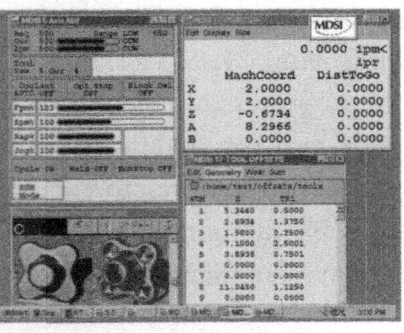

From backlash and leadscrew compensation to cutter radius compensation, OpenCNC has the tools to get the job done.

The functionality manufacturers need

OpenCNC provides the breadth and depth of machine control functionality required by manufacturers. It supports RS-274D standard part programming and is fully compatible with all major NC programming systems. Support for turrets and tool changers is standard.

OpenCNC incorporates a soft CNC and soft PLC in one application. It includes the following modules:

- human-machine interface (HMI)
- machine diagnostics tools
- soft PLC
- software motion control
- real-time data collection and distribution

It performs all of the real-time tasks such as linear, circular and helical interpolation, cutter radius compensation, velocity feed forward and 'S' curve velocity profiling in software.

Manufacturing Data Systems, Inc.
220 East Huron Street, Suite 600
Ann Arbor, MI 48104
Tel 888.OpenCNC (U.S.A. only)
Tel 734.769.9000
Fax 734.769.9112
marketing@mdsi2.com
www.mdsi2.com

Packaging for MDSI2's OpenCNC software program.

As chief technical officer and vice president for research and development, Bruce developed the world's first open CNC software for machine tools. Bruce's program could be used by any metalworking shop on a standard IBM-compatible PC with common off-the-shelf hardware. At this point, every machine tool had its own attached computer (bypassing the now antiquated punched paper tape and timesharing approach). But instead of creating custom software for each kind of machine, Bruce and Chuck's program was based on engineering mechanics about how machine parts move. "We were programming the motions, the dynamics of the machine," Bruce explains.

The result was not only a controller that could operate an NC machine but one that operated much faster. Bruce remembers piloting the OpenCNC controller on an old lathe at Brighton NC (one of MDSI's earliest customers). First, Bruce watched as the lathe operator used the company's existing controller to run the lathe. "When he got to the threading operation," Bruce remembers, "he had to reach in with a paint brush to oil the cutting tool and then reach in with a hook to remove snags of metal as they curled off the part. I watched him do this with a cup of coffee in his left hand the whole time. He picked up the can of oil, put some oil on the part, set down the oil, picked up his snag puller, pulled some snags out, took a drink of coffee, back and forth like that. Then we outfitted the lathe with my new controller. Of course, my controller's going to run it at top speed, as efficiently as possible. It was running the same part program as before, with the same operator, but when it got to the threading, he grabbed the oil paint brush in one hand and the de-snagger in the other, and he was swinging at the machine as fast as he could go. Oil, snag, oil, snag. No coffee now! He looked like a windmill, swinging his arms. When the job was done, he said, 'I never saw a machine run like this before!' We knew we had a winner."

During another test run at Great Lakes Industry in Jackson, Michigan, the operator programmed the new controller before Bruce even had a chance to show him how it worked. That's how user-friendly it was for the experienced machine operator.

Says Chuck, "I knew we had a winner in OpenCNC when Turbine Engine Components in Santa Fe Springs, California, converted every one of their 5-axis mills to our OpenCNC controller."

Chuck knew he needed a businessman to join the effort of growing MDSI2, and he convinced the founder of Great Lakes Industry, Larry Schultz (whom Chuck had known since the 1960s), to sign on as MDSI2's president. They hired application engineers in various regions to train operators, but, in Bruce's opinion, the company grew too fast. After a few years, they had run through all their funds without getting to the point of profitability. When cutbacks had to be made, even Bruce was laid off.

Says Chuck, "We attracted about seventy shareholders who invested about $14 million. But by 2002, we had not made it to break even. I was at the limit of further investment without jeopardizing my whole future life, and once I decided not to put in any more money, the rest of the investors also pulled out."

Tecumseh Products acquired MDSI2 for $.10 on the dollar, and the venture petered out. To this day, Chuck wonders if he made a terrible mistake to not find more investors to make MDSI2 a go. "Bruce's robotics software was masterful," he says. "We could have been what FANUC is today."

■　■　■

Bruce landed on his feet with a contract to adapt his software for robots in fast-paced meat-processing plants. Able to cut more than 1,000 carcasses an hour, the robots are quickly taking over a job that is notoriously dangerous for workers. "I'm replacing human beings again," says Bruce, well aware of the effects—good and bad—that his lifetime of work has had on the evolution of American industry. But, as Mike Long would put it, "Never argue with the marketplace." These men were there at the beginning of the robotic revolution and watched it unfold over decades.

In 2000, Chuck and Bruce were inducted into *American Machinist*'s Hall of Innovators, as outstanding innovators of the last decade of the twentieth century. In 2001, Bruce was honored with a distinguished alumni award from the University of Michigan College of Engineering.

In recent years, Bruce, now retired, has kept busy with many activities that make the most of his focused attention. A licensed pilot since his college days, he bought a Cessna 172 in 2004 and did all the required maintenance himself, including the work necessary to pass the annual inspection. As a member of the local chapter of the Experimental Aircraft Association (EAA), he coordinated the Young Eagles program to take kids on free airplane rides and introduce them to flying. "At least one kid I flew has become a professional pilot," Bruce says, "so I know I did some good there." Bruce also took a lead role in the EAA chapter's annual pancake breakfast, serving up to 1,000 people in just a few hours.

A lifelong acoustic guitarist, Bruce also built two guitars from scratch, under the tutelage of Chuck's neighbor in Florida (who builds guitars in his garage). "I still have both guitars," says Bruce. "One is hanging on the wall, and I keep it in tune so I can take it down and play it any time." (He also still has the first guitar he bought—in 1960.) Other retirement hobbies include swimming 1,000 meters three times a week, playing the two-person strategy game Go, and collaborating with his sister to write software that solves sudoku puzzles by mimicking the way a human would think through the solution. Bruce also meticulously cares for his 3.9-acre property west of Ann Arbor and takes pride in his two children—both successes in their careers—and his three grandchildren.

■ ■ ■

One of Chuck's many projects in more recent years was helping George Balaschak, a Florida friend, build a car from scratch. According to Chuck, George is "the finest combination of engineer, artisan, and craftsman in one human body that I have ever met." A mechanical engineer who worked on jet engines at Pratt & Whitney while restoring antique cars, George started TLC Carrossiers in 1990 to build a beautifully curved and aerodynamic automobile he called the Talbo, inspired by the Talbot-Lago, a 1930s French car designed by Figoni and Falaschi.

1996 Talbo. Photo: George Balaschak.

When Chuck saw George making one of the first Talbo chassis, he introduced George to computers and lent him his PC with the first copy of AutoCAD installed. (The only plotter driver available with this release of AutoCAD was for the MDSI plotter.) George put all the drawings for the pieces of the chassis in AutoCAD. Then they found a local shop with a CNC laser cutter, and George had the pieces for the next chassis cut from the AutoCAD drawings. As he assembled the next chassis, every time he had to change the size of a piece a little bit to make it fit better, he modified the drawing. Thus, every chassis after that got better and better.

Once George was into CAD, the next thing he did was design a fixture to aid in assembling the chassis. He was using the same techniques he had just learned using CAD and laser cutters. For Chuck, this was the beginning of the realization of his dream from the early days of MDSI and the DESIGN system.

By 1995, George and his Talbo had attracted the attention of Frank M. Rinderknecht of Rinspeed, Inc., a Swiss concept car builder. Rinderknecht commissioned George to build a customized Talbo, named "Rinspeed Yello Talbo," for the 1996 Geneva International Motor Show in Switzerland. Then Rinderknecht came back with a request to build a car styled after a Barney Oldfield open-wheel racer for the 1997 Motor Show.

At this point, George made an inquiry to Parametric Technology about hiring some help to use Pro/ENGINEER. He learned that tech consulting was available for $1,000 per day. George said, "Let me think about it."

The following Monday, a Pro/E tech guy arrived at Balaschak's door. Chuck salutes George for not losing his cool since he had not authorized this $1,000-a-day support person. The two men started working together, taking turns in the "driver's seat," and the day went amazingly well. George agreed to continue for the rest of the week, at which point they had the beginning of the vehicle that Rinderknecht wanted. George could already see what was shaping up to be a car built like an airplane with ribs and stringers.

When he got the body the way he wanted it, he asked the Pro/E expert to return and help him get all the notches in the correct locations so the ribs and stringers could be put together. When that was complete, he released the cutting of the 3/16" sheet aluminum, and as pieces were returned to him, they were assembled. When that was finished, the joints of the ribs and stringers were welded together.

George hired a couple of skilled European technicians to shape the aluminum skin for the car. Rinderknecht showed the car, named "Rinspeed Mono Ego," at the 1997 Geneva Motor Show.

"Rinspeed Mono Ego" in process. Photo: George Balaschak.

Later in 1997, Rinderknecht returned to George. This time he wanted an eyebrow-raising supercar to be named "Rinspeed E-Go Rocket." George was working on drawings for the car when he and Chuck bought a used NC milling machine for the Talbo shop.

The machine had an 84" X-travel and 44" Y-travel, but the spindle and Z-axis were pretty much useless because they were in such bad shape. Chuck designed, using AutoCAD, a new Z-axis with two rotary axes (A and C) on the end. That was a short trip back to Chuck's machine tool days at Buhr.

NC milling machine in George Balaschak's shop.

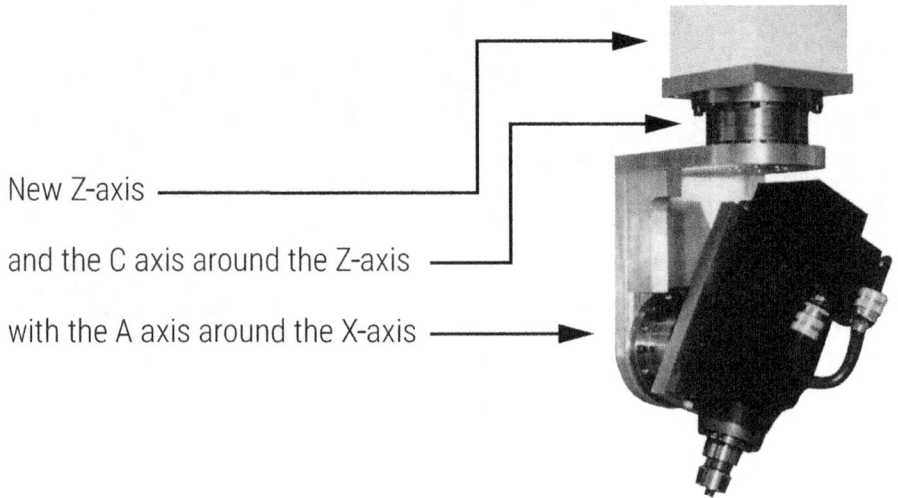

New Z-axis

and the C axis around the Z-axis

with the A axis around the X-axis

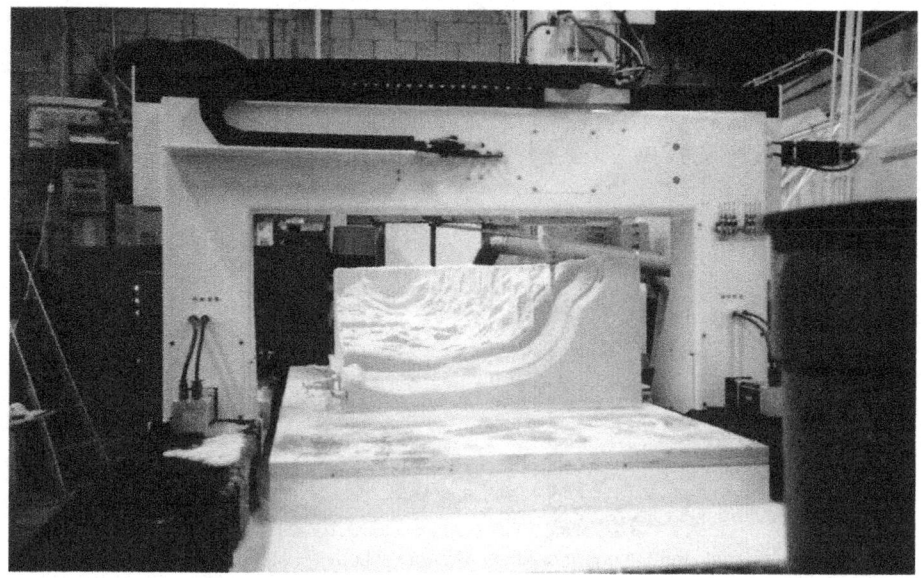

Cutting the foam mold for one quarter of the body of the Rinspeed E-Go Rocket.

Chuck and George then installed OpenCNC on that machine and got it running in the last week of December 1997. The OpenCNC software developed by MDSI2 worked right off the bat—a real tribute to Bruce Nourse. Because it had taken a while to make the NC machine usable, time was running short to complete the supercar before it had to be shipped for the next International Motor Show.

Using Pro/ENGINEER to generate the tool paths, George and Chuck made twenty-two molds from huge blocks of foam.

After successfully cutting the first quarter-body mold, they were about to cut the second when Chuck looked at George and said, "That seemed flawless! I'm not going to stand here for several hours and watch that machine make foam dust. I'm going home to get a good night's sleep."

On the way out the door, George started to set the burglar alarm. "Stop!" said Chuck. "The machine motion will immediately set off the alarm." So they locked the door and left. They returned the next morning to a huge pile of foam dust that was hiding another beautiful quarter-body mold, with every detail finished to perfection.

George then made the fiberglass parts in those foam molds and assembled the pieces to the already complete chassis of the car. He test

Finishing the foam mold for one quarter of the E-Go Rocket body.

1998 "Rinspeed E-Go Rocket."

drove the finished E-Go Rocket around Riviera Beach, Florida, with a blue-haired model in the driver seat, and then shipped the car to Switzerland just in time for the Geneva International Motor Show in March 1998.

In sum, George was able to make twenty-two molds, make the fiberglass parts in those molds, and assemble the whole car in about nine weeks!

Chuck looks at that project as the final step in proving that his original goal for DESIGN was the correct path—"from CAD to manufactured parts using CNC." Says Chuck, "I give special thanks to Don Feddersen and Sam Geisberg for Parametric Technology and Pro/ENGINEER. I owe George Balaschak a lot for helping me see the realization of my very earliest vision that goes all the way back to my days at Buhr Machine Tool Company in 1965."

In recent years, Chuck's favorite retirement activity is supporting the undergraduate students on U-M's Solar Car Team. With its lessons in engineering as well as teamwork, leadership, and project management, the solar car program builds the kinds of U-M graduates who go on to make their mark in various industries and business roles. Chuck has mentored students from many disciplines as they design and build a solar car to compete in the North American Solar Car Challenge. He is continually encouraging students to get into the metal shop themselves to make the parts they need.

A frequent U.S. champion, the U-M Solar Car Team almost always competes in the World Solar Car Challenge held in Australia. Chuck has been there eleven times to travel the race route from Darwin to Adelaide with the team.

Now in their eighties, Chuck and Ann have targeted much of their philanthropy to the University of Michigan, including the Harold A. Furlong Professorship in Women's Health in the U-M Health Center's Department of Obstetrics and Gynecology (named after the Pontiac doctor who mentored Chuck in his teen years) and the Frances B. Furlong Scholarship in the School of Music, Theatre & Dance.

Chuck (on left) with the 2017 University of Michigan Solar Car Team in Adelaide, Australia.

· · ·

After selling MDSI and working just two years for Schlumberger in New York, Ken Stephanz returned to Ann Arbor and struck out on his own as a business and management consultant. He assisted several companies with successful turnarounds. For a few years, he was a silent partner in a natural gas drilling business in the Southwest. "It was rewarding," says Ken, "but not nearly as much fun as MDSI."

Reflecting, Ken adds, "I was extremely reluctant for MDSI to be acquired or merged with anybody. The subsequent, actual demise of MDSI followed the often-trod path of other independent, sparkling companies that are acquired or merged, wherein the key players leave, sales sag, the whole equation eventually falls apart, and nobody wins. It is a bittersweet pill to swallow and a song that is sung much too often. At the same time, the original investors needed a return on their investment, and 'going public' was a sure sign that this scenario would be the final outcome. For some of us, it was very painful."

Ken's wisdom and years of experience in all phases of business development were invaluable to the companies and nonprofits he assisted in the U.S. and internationally. He helped found two Michigan-based nonprofit economic development organizations: the Industrial Technology Institute and the Center for Entrepreneurship at Eastern Michigan University. Ken served on so many boards simultaneously that he once figured out he had accumulated 140 years of board service.

In the 1990s, Ken traveled the world to provide economic development advice to emerging governments. He made trips to Thailand, Vietnam, Ukraine, Zimbabwe, Nigeria, Ghana, Mozambique, South Africa, and Mexico, often working on challenging projects to combat poverty. "We have no idea how hard people have it," Ken says.

As time allowed, he still enjoyed piloting small aircraft and singing music, and playing golf on some of the country's best courses. When he moved with his wife, Edie, to Florida in 1995, he also took up SCUBA diving and deep-sea fishing.

For the next decade, he was a certified business analyst in Florida for the nationwide, federally funded Small Business Development Center Program, providing free assistance to entrepreneurs starting and growing their own small businesses. The SBDC recognized Ken with multiple awards for counseling excellence. "I was named the best counselor of the 175 business analysts in Florida in eight of the nine years I was there." Even in his eighties, he still knew how to nurture people toward success.

After thirty-nine years of marriage, Ken lost Edie to Alzheimer's in 2010. The cost of her care in those later years depleted Ken's resources but not his determination. At ninety-four, he still lives independently in his own home with a beloved canine companion—a Doberman-mix named Lexi. "My body has betrayed me," Ken says, referring to chronic pain, "but my mind is better than ever. The only thing I have left at this age is my integrity, but I feel strongly about that." Ken lives only about an hour north of Chuck, and the two old colleagues occasionally get together.

■　■　■

When Mike Long left MDSI, he was looking "to have a CEO experience." He moved with his family to northern California, where he worked as a consultant to several venture capital firms, vetting business plans and helping start-ups. He served as president and CEO of an electronics company in Sunnyvale until he sold that company. In 1997, he bought NDT Laboratories, Inc.—an industrial testing lab—which he owned and ran for the next eighteen years. Says Mike, "We did a lot of work analyzing electronic components and aerospace parts and missile parts, what we call highly engineered parts. That was a really good business."

Mike sold it in 2015 and retired. He's still in California today, playing golf every chance he gets.

■　■　■

When "Van" Van Bemden was laid off by Schlumberger, he went to work as the primary part programmer for one of MDSI's earliest customers, Brighton NC. In 1995, he left that job to work for a year with Focus Hope in Detroit. The nationally recognized nonprofit offers multiple services to overcome racism, poverty, and injustice, including a robust job training and apprenticeship program.

Van became a supervisor in Focus Hope's NC machine department, which prepared people for jobs with the Big Three automakers. "Many of the larger NC machine tool manufacturers had their latest NC machines in our plant," Van remembers. The trainees, mostly young African Americans, operated the NC machines for six hours a shift, creating real parts for customers, and then they moved to the classroom to learn some shop math, communications, and technical skills like how to read a blueprint.

In 1997, Van retired and moved to Colorado with his wife, Brenda, to live near their grandchildren. He built houses for Habitat for Humanity, skied a lot, and started hiking Colorado's 14,000-foot mountains. He hiked his first "14er" at age sixty-three and eventually hiked all fifty-five peaks, nineteen of them in a single summer.

Through his church, Van also got involved in mission work in Santa Cruz, Bolivia, helping to design and build a church and orphanage there. When the need for real architectural blueprints became obvious, Van put his CAD/CAM experience to work, learned to use professional architectural software (provided to him at no expense by a former MDSI colleague), and became the mission's go-to guy for building design. This led to a similar volunteer role for the construction of an orphanage in Motipur, India. Van traveled there three times to assist with construction.

Van is, of course, no stranger to international travel after his many years with MDSI. He still tells people that he spent his career at "the IBM of numerical control programming."

■ ■ ■

On June 15, 2018, Chuck and Ken and Bruce and Van and more than one hundred former employees of MDSI gathered in Ann Arbor for a reunion. They toured the company's former headquarters on Plymouth

Road—now the home of U-M's Information and Technology Services—and then enjoyed a long dinner filled with storytelling.

As it happened, just a couple of miles down the road, on U-M's North Campus, construction was starting on the Ford Robotics building for the development of all sorts of robotic technology. It seemed a fitting convergence in time to welcome back to Ann Arbor the early pioneers in numerically controlled machine tools just as U-M and Ford were committing new resources to the robotics made possible by those pioneers.

Chuck, who organized the special event, called it the "50th Reunion," because it was held exactly fifty years to the day since Chuck hired Van to join his endeavor. Bob Pavey was there, representing Morgenthaler Associates, and Comshare's Bob Guise came, and Seth Powsner, and Al Kortesoja.

Alfred Vieth and his wife, Jutta, flew in from Germany and reconnected with Yoshi Taguchi, who flew in from Tokyo. The two reminisced and laughed about their MDSI training days when they saw a streaker in Ann Arbor on their way for late-night pizza.

After he left Schlumberger, Yoshi went to work for Electronic Data Systems and started the EDS–Japan office as a vice president. It was Van who told Yoshi about the reunion. "I immediately decided it's a right decision to come and see all of you," he said. "I was really excited."

Alfred's career after MDSI continued in the field of 3-D CAD and programs for 4-axis and 5-axis machines. He remembers, "When people came to demonstrate new systems, they were mostly ex-colleagues from MDSI. I owe MDSI a lot for helping me in my career."

Others at the reunion told similar stories of running into former MDSI folks at various jobs throughout the world, or of deciding to hire someone when they saw MDSI on the applicant's resume. As reunion attendees took turns at the microphone to reminisce, there were geeked-out reminders of programming feats, howls of laughter at travel adventures gone awry, and rounds of applause in agreement of the amazing experience they all had shared.

"The ambiance of MDSI was unique," Ken remembered. "Happy, hardworking, fun, with a sense of 'we can do anything.' And we did!

The feeling of success, quality, accomplishment, and pride permeated everything in the MDSI universe. It was contagious and wonderful. We lived in a magic, Camelot time." Ken told his former colleagues, "You have my eternal gratitude for allowing me to be part of this incredible journey."

Chuck beamed with pride throughout the entire event. With each story he told, he conveyed a sense of unremitting awe at how it all worked out. "I don't know how I was so lucky. I worked with so many super people. And we can look back on so many examples of how our work was on the leading edge of technology. I can't think of any other time in the history of man that I would rather be alive than right now."

ACKNOWLEDGMENTS

Bringing this book to publication has truly been a team effort.

My greatest appreciation goes to Stephanie Kadel Taras, PhD, of TimePieces Personal Biographies, LLC. Stephanie used transcriptions from videos taken at the MDSI 50th Reunion in June 2018. She also collected writings and interviewed key MDSI people in person and by phone. She grabbed those stories and accounts, incorporated her extensive research, and knitted it all into a book. No one could have written it better.

I am most grateful for the MDSI people who spent extensive time with Stephanie helping her flesh out the story; their remembrances are the backbone of the book. Included here are Ken Stephanz, Bruce Nourse, Cai Raber, Chet Fleszar, and Mike Long. Others who contributed shorter interviews were Tim and Tom Clausnitzer, Judy Foster Leverett, Teresa Killeen, Seth Powsner, Urbanes Van Bemden, and Bob Pavey.

My appreciation to all the people who didn't know they would contribute to a book when they wrote memories, spoke at the 50th Anniversary dinner on Friday, and shared stories during the picnic on Saturday. Their recollections supplied details important to making the story complete.

Thanks to both Bruce Nourse and Urbanes Van Bemden for providing photos and to Bruce for also adding archival materials.

As the book progressed, several people assisted in reading and editing. I give special thanks to Ken Stephanz for his many thoughtful suggestions. I want to also thank Mike Long, Bruce Nourse, Seth Powsner, Urbanes Van Bemden, and MDSI board member Bob Pavey, who read the manuscript along the way and offered their advice.

Those transcriptions from videos, referred to above, are the excellent work of Evan Dougherty and his wife, Ginny, who not only videotaped the reunion dinner speakers on Friday evening, but also the stories collected during the picnic on Saturday. I much appreciate Evan's proficiency, which I first became aware of when he videotaped the U-M

Solar Car Team in Australia in 2011.

Thanks to Gary Morgenthaler for his insightful quote regarding MDSI's pioneering role in bringing automation to manufacturing.

I am so impressed by the fantastic cover design and page layout by Lisa Armstrong, graphic designer, of Ajuga, Inc. Thanks to Lisa for capturing the story's essence in the cover.

Once Stephanie had a manuscript, she suggested we turn it over to Rae Jean Sielen and her assistant, Andrew Rorabaugh, of Populore Publishing Company, for their expert fine tuning. I give them a huge amount of praise for their mastery. Without them this book would just be like the egg that didn't hatch.

I want to express my gratitude to Ed Downing; his help and patience were responsible for my evolution from a mechanical engineer to a computer techie.

I am indebted to Bob Guise who, after a coincidental sidewalk meeting, took a chance on me when I only had an idea.

Special thanks to Larry Schultz who was one of the very first users of COMPACT while still a student at Michigan, and who also initiated our venture into greater complexities at Great Lakes Industries in Jackson, Michigan.

Ken Stephanz, CEO extraordinaire, has always attained the top of whatever he's undertaken. No one else could have led MDSI as well as Ken, and he has my utmost appreciation.

I was so fortunate to have my excellent R&D team of Bruce Nourse, Seth Powsner, Don Willan, Don Colley, Dave Jenson, Dave Hinckley, and Drake Fink who, among their many accomplishments, spent forty-nine man-days compressing the code in the "A" page so we could accommodate Mike Long's realization that the world was going metric. I can't give them enough praise for their coding skills.

MDSI would not have survived without Betty Ruddy, Carol Guttman, Judy Foster Leverett, and others like them, who were really in charge of their respective executives. I applaud them for holding everything together.

Thanks to Mike Levine, whose CRT and keyboard helped launch the ST-1 as an early personal computer.

I am grateful to David Morgenthaler who, as a young venture capitalist, decided to take the risk of backing a fledgling company in an emerging field.

My particular thanks to Bruce Nourse for his lifelong support and friendship. Bruce used his brilliant programming to conquer every challenge.

I want to recognize all the employees who weren't mentioned in the book. There just isn't space here to include all of the 1872 incredible people who were part of the story. Their dedication and expertise helped make MDSI one of the most successful businesses in Ann Arbor at the time.

I give my applause to the wives, husbands, and significant others who held down the fort while their partner was late for dinner, out of town, or missed an event. Their backing was crucial to the success of MDSI.

George Balaschak gets credit for proving my concept for DESIGN— by using computer-aided design software to design a car body, and a 5-axis numerically controlled machine tool to make it.

Thanks to our three children for their encouragement to put some of my memories and the MDSI story in writing. Linda recalls working for Judy Foster Leverett the summer of 1972. Beth remembers the summers of 1978 and '79 when she worked in documentation and marketing. And when Stuart was in his early teens, and construction was starting on one of the buildings, we climbed up on a monster earthmover, and I let him drive. Dick Stitt chewed me out for that.

My highest appreciation goes to my wonderful wife, Ann, for sixty-five years of love and affection, and all her effort in making this book come to fruition. I could never have done it without her.

NOTES

1   Leland, Wilfred C., and Minnie Dubbs Millbrook. *Master of Precision: Henry M. Leland*. Detroit: Wayne State University Press, 1966.

2   See relevant history at http://www.lathes.co.uk/devlieg/ and http://www.lathes.co.uk/devlieg-jigmil-type-b/. Accessed 09/22/2020.

3   Wikipedia entry for SDS 940. https://en.wikipedia.org/wiki/SDS_940. Accessed 09/22/2020.

    "SDS 940 Timesharing Computer Control Panel (1966)." Artifact document from Computer History Museum. https://www.computerhistory.org/collections/catalog/102762446. Accessed 09/22/2020.

4   Crandall, Rick. "Advent of Graphical Executive Information Systems." Artifact document from Computer History Museum, 2009. https://www.computerhistory.org/collections/catalog/102762441. Accessed 09/22/2020.

    Oral history interview of Rick Crandall by Paul Ceruzzi, May 3, 2002, Washington, D.C. Minneapolis: University of Minnesota, Charles Babbage Institute, Center for the History of Information Processing.

    "Timesharing/Remote Processing Services Workshop: Session 2: Formation of TS/RPS companies." Moderator: Burton Grad. Mountain View, CA: Computer History Museum, June 2, 2009.

5   Brochure from University of Michigan Manufacturing Workshop, November 2 and 3, 1967.

6   Carnahan, Brice, H.A. Luther, and James O. Wilkes. *Digital Computing and Numerical Methods*. New York: Wiley, 1964.

7   Hutchins, Charles S. "Time-Sharing Advances in Computer-Assisted Parts Programming and Their Application to Direct Numerical Control." *NC Scene*, October 1971.

8   "Manufacturing Data Systems, Incorporated." MDSI internal document, October 29, 1970.

9   Ibid.

10 Bendix Systems Division, Ann Arbor, Michigan, magazine advertisements, 1958–1960, posted on Flickr.com. Collection of Wystan Stevens.

    McLeister, Dan. "Bendix May Sell All of Its 43-Acre Site." *Ann Arbor News*, October 16, 1974.

    Slagter, Martin. "Apollo 11 Mission Had Deep Ann Arbor Ties." *MLive*, July 18, 2019. https://www.mlive.com/news/g66l-2019/07/a5482cdbf95647/apollo-11-mission-had-deep-ann-arbor-ties.html. Accessed 09/22/2020.

11 "Time-sharing for numerical control." *Com-Share News*, Fall 1970.

12 Claude Wilson obituary, *Sun Sentinel*, January 25, 2001.

13 Mike, Elizabeth. "N/C Machine Tapes Prepared Faster by New Method: MDSI." *Metalworking News*, August 18, 1969.

14 Ibid.

15 "Dial for Data Processing." *American Machinist*, November 3, 1969.

16 Stephanz, Kenneth R. "Stockholders' Annual Meeting: President's Message," December 18, 1970.

17 Time-sharing for Numerical Control." *Com-Share News*, Fall 1970.

18 Navellier, Louis. "(Back to) Earth Day: The Tech-Stock Crash of 1970." *Nasdaq News*, April 23, 2010. https://www.nasdaq.com/articles/back-earth-day-tech-stock-crash-1970-2010-04-23-0. Accessed 09/22/2020.

19 Mari, Albert. "MDSI COMPACT II System Extended Into Turning." *Metalworking News*, September 13, 1971.

20 Mari, Albert. "Manufacturing Data Pushes Ahead; COMPACT II Now 'Speaks' Fluent Metric." *Metalworking News*, March 13, 1972.

21 "MDSI: Shops on N/C Bandwagon." *AMM/MN*, February 5, 1973.

22 Mari, Albert. "MDSI COMPACT II System Extended Into Turning." *Metalworking News*, September 13, 1971.

23 Mari, Albert. "Manufacturing Data Pushes Ahead; COMPACT II Now 'Speaks' Fluent Metric." *Metalworking News*, March 13, 1972.

24 Untitled article in *Ann Arbor News*, May 28, 1971.

    Mari, Albert. "MDSI COMPACT II System Extended Into Turning." *Metalworking News*, September 13, 1971.

25 "Manufacturing Data Systems, Inc. Announces Significant Product Expansion," MDSI internal document, June 5, 1974.

    Truax, Julie B. "MDSI." Graduate student research paper, based on interviews with Charles S. Hutchins and others, December 11, 1981.

26 Source for the 1970 cost of the TI-980: https://news.google.com/newspapers? nid=849& dat=19700527&id=Pf4vAAAAIBAJ&sjid=sU4DAAAAIBAJ&pg= 4568,5593679. Accessed 09/22/2020.

27 "MDSI Chronological History." MDSI internal document, October 1981.

28 Ibid.

29 Much of this section comes from "MDSI Building Fact Sheet," June 30, 1980.

30 Klein, Pamela. "MDSI adding $9 million building." *Ann Arbor News*, May 7, 1980.

31 "Manufacturing Data Systems, Inc.: Ann Arbor's Worldwide Leader in Computer-Assisted Numerical Control." *Ann Arbor Scene Magazine*, 9(3), Fall 1980.

32 "MDSI Chronological History." MDSI internal document, October 1981.

33 "Manufacturing Data Systems, Inc.: Ann Arbor's Worldwide Leader in Computer-Assisted Numerical Control." *Ann Arbor Scene Magazine*, 9(3), Fall 1980.

34 "MDSI Financial Statistics." MDSI internal document.

35 Spalding, Oakes A., Jr. "100 Emerging Growth Stocks in High Technology Industries." Comments from Adams, Harkness & Hill, Inc., Boston, Massachusetts, August 18, 1980.

36 *1981 Schlumberger Annual Report.*

37 "MDSI Financial Statistics." MDSI internal document.

38 Benner, Katie. "David T. Morgenthaler, Who Shaped Venture Capitalism, Dies at 96." *New York Times*, June 21, 2016.

39 This single sheet from a larger report does not provide citation information, but the content, the writing style, the font, and the margins suggest it is very likely another report from Adams, Harkness & Hill, Inc., Boston, Massachusetts (see citation above). A handwritten date cites January 1981.

40 MDSI internal document of corporate statistics as of January 1, 1981.

"MDSI Chronological History." MDSI internal document, October 1981.

41 Klein, Pamela. "'Good things' ahead for MDSI." *Ann Arbor News*, January 14, 1982.

42 *1985 Schlumberger Annual Report.*

43 Moilanen, Kathy Ann. "Firm revives Applicon name, plans to go public." *Ann Arbor News*, September 16, 1992.

44 Morgan, Mary. "U-M Purchases Buildings, Land of Arbor Lake Complex." *Ann Arbor News*, October 5, 1997. In 1997, Domino's Farms bought forty-eight acres of undeveloped land from Schlumberger, and U-M bought the three buildings and twenty-two undeveloped acres.

INDEX

Note: Photographs and illustrations are indicated by italicized page numbers.

right hand—to start the spindle going in a clockwise direction), 31

Ruddy, Elizabeth "Betty," 109, *110*, 117, 125, 127, 136, 146, 160, 162, 222

S

S&H Green Stamps, 136

Samford, John, 196–197, 223

Sayler, John, 223

SBQ command, 181, 190

Schlumberger
  acquires Applicon, 221
  acquires MDSI, xi, 216–220, *219*
  Computer Aided Systems, 221, 226
  as customer of MDSI, 152
  fallout from acquisition of MDSI, 221–230
  merges MDSI into Applicon, 229

Schlumberger Technologies, 226
  sells MDSI/Applicon to Gores Enterprise, 229
  use of distributors vs. subsidiaries, 227

Schofield, Keith, *170*

Schroedel, Gunther, *170*

Schultz, Larry, 62, 236

Scientific Data Systems (SDS)
  sale of company to Xerox, 111
  SDS 930 computer, 52–53, 54
  SDS 940 computer, 53, 54, *55*, 57, 59, 111

Searle Pharmaceutical, 163

Selig, Juergen, *162*, 166, *170*

Sentry, 226

Sheridan, Richard, 223–225

Shingler, George, *211*

Shuman, John, 191

SIC codes, 153

Siegel, Kip, 33

Sigmund Blum & Associates, 194

Silfverling, Olof, *162*

Simon, George, 129, 193

Sinclair, John, *170*

single-pass processors, 76

Siri, 106

sleep, revelation of solutions during, 33

Small Business Development Centers (SBDC), 245

Smart Terminal 1 (ST-1), 188–190

Smith, Michael, *170*

Society of Automotive Engineers conference (1968), 62–63

Society of Manufacturing Engineers (journal), 87

Software Algorithms, Inc. *See* MDSI2

Spencer, Jim, 152, 153

SPLIT (Sundstrand Programming Language Internally Translated), 31, 34, 36, 44, 59, 60–61, 88, 125

spreadsheets, 133–136

Stancato, Frank, 149

Stephanz, Edith Mary, 95, 102, 245

Stephanz, Francis, 95–96

Stephanz, Glenn, 95

Stephanz, Harold, 95

Stephanz, Kenneth R.
  at annual meetings, 173
  background and early career, 92–93, 95–101
  exposure to NC, 102
  as founder of MDSI, ix–xiii, 102–117
  as founder of nonprofit economic development organizations, 244
  as international business consultant, 244

Made in the USA
Monee, IL
26 October 2021